FAST TRACK I.T. JOURNEY

How To Move From Supplier To Partner

Alok Ranjan Tripathy

FASTTRACK I.T. Journey – How to move from Supplier to Partner

Copyright @ 2018 by Alok R Tripathy

ISBN-13: 978-1729621950
ISBN-10: 1729621953

No part of the publication may be reproduced, stored in a retrieval system or transmitted in any form or by any means, electronic, mechanical, photocopying, recording, scanning or otherwise.

All rights reserved, including the right to reproduce this book or portions thereof in any form whatsoever.

Published by:
10-10-10 Publishing
455 Apple Creek Boulevard
Suite 200
Markham, Ontario
L3R 9X7

First 10-10-10 Publishing paperback edition November 2018

Table of Contents

Foreword	vii
Acknowledgements	ix
Chapter 1 – Future is Beautiful	1
Chapter 2 – You Have Been Faanged	7
Chapter 3 – The Conundrum – Service Provider	13
Chapter 4 – Transformation Journey	19
Chapter 5 – Customer Centricity	25
Chapter 6 – Communication	33
Chapter 7 – Creativity	41
Chapter 8 – Commitment	49
Chapter 9 – Capability	57
Chapter 10 – Collaboration	63
Chapter 11 – Continuity	69
Chapter 12 – Sharing	77
Chapter 13 – Engage	85
Chapter 14 – Assessment	93
Chapter 15 – Leading or Being Led	99
About the Author	109

*I would like to dedicate this book to my dad, Padma Charan Tripathy, and mom, Anita Tripathy.
Thanks for pushing me to new frontiers all the time, and for providing me all the support to be where I am today.
Love you Baba and Mimi!!*

Foreword

FastTrack Your I.T. Journey, by Mr. Alok R. Tripathy, is intended to help you decode the definition, and nuances in roles and responsibilities, of two key players in the I.T. journey.

This book will give you the perspective of both supplier and buyer in I.T. services, and will also explain the importance of the partnership in catapulting your productivity exponentially.

Do you know how much productivity wastage can be recovered, and therefore how much benefit can be shared back with your buyers? This book brings out some uncanny parlance form other industries in terms of practice and mature behavior, which can be leveraged to mutual benefit to the partnerships between buyer and supplier.

The findings in the book will force you to think differently, whether you are an I.T. buyer or supplier. Taking some useful small steps will result in a conducive environment for the workforce on both the sides to co-exist and perform as a powerful team for greater goods.

Alok R Tripathy is a successful I.T. consultant, and has worked across Europe for the last 20 years. He has grown along with the I.T. uprising, and he has a deeper understanding of how the industry has been shaped, from monolithic mainframe boxes to modular customer centric crowdsourcing platforms.

I first met Alok in early 2018 and was impressed with his clarity of thought. I found him immensely talented and knowledgeable in the minute details of the I.T. service industry. Alok is very passionate about his job, and is a man of integrity.

This book is an amalgamation of his experiences, success stories in other industries, and the lessons he learned from failures in the marketplace. Alok has been successful in deciphering a complex problem and coming up with simple yet relevant steps to challenge the status quo for the better.

I am sure you will find this book informative, and I cannot wait to see you leveraging a few nuggets of wisdom in your strategy and goals in your workplace.

Let's go for a new beginning!!

Raymond Aaron
New York Times Bestselling Author

Acknowledgements

I am grateful to my wife, Suravi Mishra, who has been my strength and support. She has been with me during my ups and downs. I always feel she is a much better manager than me in real life. She is the spiritual guru and yoga instructor at home. I envy how she can manage so many roles, with utmost ease and a beautiful smile.

I would like to thank my sisters, Shivani Devta and Himani Mohapatra. They have been my critics and challengers throughout my life. I admire my sisters for their courage and positivity, even in adversity. They always stood firm to shoulder the family responsibility, so that I could continue pursuing my dream.

Thank you, my two little angels, Sheetal and Komal. I admire Sheetal's eye for detail, and her focus on whatever she does. She is a task master, and I wish her all the happiness and success in her life. Komal is a fearless fighter and takes on whatever life throws at her. She is the baby of the family and will always remain the entertainer for us. We love but are sometimes annoyed at her "I want it, I got it" attitude.

Special mention to my best buddy, my cousin, Siddhartha Satpathy. He knows most of my secrets!! He is my brother, friend, and partner in crime till today. I admire his resilience and confidence in whatever he does, and I am proud to have him in my life!!

I would like to call out the rest of my close cousins—Rachita, Sucheta, Sharmistha, Budhi, Tiki, Tata, Pipi, Reena, Rinku, Biswajit, Manoj, Rajat, Saroj, Arun, Shakti, Babu, Parth, Swagat, and Tithankar—in my growing up years. The big boys and girls have always guided, helped, and inspired me throughout my childhood. I cherish the memories of the growing phase of the younger ones; they often pleasantly surprised me with the kind of achievements they have attained in life. I thank all of them for being there for me whenever I needed them.

The acknowledgement section would be incomplete without thanking Uncle Ravi Tripathy. My childhood would not have been so memorable without him being in my life. He always treated me as friend and, till date, I love him the most, though it's difficult to say in as many words when we meet. Hence, this is my gratitude to him for the good times he has given to me.

I would like to thank my first teacher, Dhoba Pradhan. All my success in early years would not have been possible without his enforcement of strict discipline, both in academia and extra-curricular activities.

Let me take this opportunity to thank all the industry leaders with whom I have been brought up. I cannot name all of them, given that I am going strong, even after two decades in my esteem organisation, and some of them have moved out. To start with—BG Srinivas, Mohit Joshi, Rajesh Krishnamurthy, Ravi Kumar, Ajay Vij, Sanjay Jalona, Praveen Kumar, Kannan Amaresh, Prakash Chellam, Pascal Beignon, Jitin Goyal, Praveen Kumar, Atul Soneja, Anup Upadhyay, Ramesh Amancharla, Karthik Ranganathan, Roshan Shetty, Amit Vasudeva, Harpreet Singh, Shripad Shanbag, Guru Murthy(Gurums), Mark Holden, Sunil Khurdi, Vivek Rana, Mark Bryant, Rick Eager, Sushil Kumar, Sagi Ramachandra Raju, Sembi Sahab, Kaustub Nadgir, and Milind Pol—I have been indebted to all these folks, who have always guided me on what's right for me. Also, they have applauded my courage and conviction throughout.

Special thanks to each and every colleague who has been my friend, advisor, and critic: Saket Singh, Vishal Dutta, Sujeet Singh, Rajesh Sadashivan, Pragna Sen, Vinod Balakrishnan, Nagaraj Nanjundram, Sam Prashanth, Arindam Ghosh, Puneet Shukla, Saurabh Bagrodia, Anand Prahladsingh, and Chandra. Without their support and competitiveness, this book would not have been possible.

In the last 20 years, I was fortunate to get an opportunity to work with some of the best minds across countries. Without their contribution to my experience, this book would not have been complete. So I would like to thank Ian Scott, Sally Robson, Mark Diamond, Mark Bryant, Steve Croke, Simon Corcorron, Benny Higgins, Peter Dingle, Andrew Berry, Monique Riddle, Monique Heuben, Rick Gray, Stephen Sorin, David Trahand, Maxim Gaudin, Katrina Karolinen,

Jacqui MacDonald, Barbara Moldenhuer, Sundeep Goel, Douglas Mcewan, Nico Lassing, Chris Booth, Phil Brandom, Rosa Henriques, Peter Quist, Raymond, Edwin, Rick Eager, Declan O'gormon, and Stephen Aitken.

Special mention to two of my customers who always challenged me to do things today if planned for tomorrow: Mike Kennelly and Will Furneaux. I like and respect their drive and passion to make things happen in the most complicated environments.

I have to mention these three names: Boogie, Arlene, and Matty. For the last 11 years, I have never missed your show in the morning on Forth1 Radio. I admire you folks for the fearlessness and the humour you bring in every day, year after year. I can safely say, "You made my day." I still remember the day when you mispronounced my daughter's name as *Kamal* (for Komal), for "Joke – Kids in Car." Don't worry; we, with our Indian accent, do the same for your name: *Boogy* for Boogee.

Due to my papa's job as a civil engineer, I had to change five schools up till my senior years. No complaints!! Thanks to technology, I am in touch with a couple of school groups on social media. I would like to recognise their role in my life while I was growing up. Thanks to my real competitors in school: Babita Samal (toughest competitor), Priyambada Roy, Santanu Maharana(Sana, the glue of the group), Basudev Purohit, Namita Biswal, Swati, Nalini, and Karunakar Raut.

My special gratitude goes towards three key inspirational personalities, for many in India and across the globe: Finance Minister, Dr. Manmohan Singh, who was the chief architect of modern India in the early 90s; secondly, the current prime minister, Mr. Narendra Modi, the pilot in chief for taking India to the next level in the socio-economic-political ladder; and lastly, the man who has re-defined suave, proficiency, and eloquence when he speaks, Dr. Sashi Tharoor. Without his advice, I might not have written this book. It all changed when I was fortunate to have lunch with him, on his trip to Edinburgh.

Life has always been kind to me, as I had one of the best upbringings—though I never used to appreciate it when I was young. Thanks to Dr. Manmohan Singh, the erstwhile finance minister in India, who opened the door in 1991, in the name of economic liberalization. Then there were 3 distinguished pioneers who latched onto the initiative headed by Mr. Narayan Murthy, fondly known as NRN. The other two partners in crime, with NRN, were Mr. Ratan Tata and Mr. Ajim Premji. I was blessed to work in the era of so many great visionaries from India, like Mr. Nandan Nilekeni, Mr. Chandra, and Mr. Shiv Nadar.

Thanks to Theresa May and Nicola Strugen for showing extreme resilience during the challenging times with Scottish Referendum and Brexit talks in the last couple of years. It reflects the true character of a leader when the going is tough.

I would like to take this opportunity to express my gratitude towards Her Majesty, Queen Elizabeth, for her longevity and essential role in heading the commonwealth agenda efficiently. The significance of holding the group is paramount where the world is drifting apart due to ideological differences.

The book would be incomplete without a special mention to my badminton club (Lothian and Border Police Club) in Edinburgh: John, Gary, Margaret, and Irene. Thanks to you all for keeping me sharp on the court all the time.

My friends in my social circle— Gautam, Mitra, Manas, Sheela, Prabhat, Debabrata, Dilip, Nilkantha, Santoshini, Sarita, Anandita, Bhajaman, Bishnu Mausa, Anil and Suravi Chugh, and Manjari and Rajnish Singh—have always stood by us, and I would like to share my appreciation for everything they have done for me.

"Love All, Serve All": I am amazed by the level of customer centricity shown by all the staff of Hard Rock Café in Edinburgh. I have learned a lot from their professionalism.

My sincere gratitude towards, Raymond Aaron, who inspired me to write a book, the very first time I heard him. I would like to thank Chinmai Swamy, my Book Architect and Lisa Browning, who went through the painstaking job of editing and formatting for my book.

Last but not least, I would like to share my appreciation for Lubeyen, a 21-year-old student from Bulgaria, who I had met in a restaurant. I was inspired by his dedication to do something great in quantum physics and planetary movement. Later, he received an apprenticeship from Google. I salute his dedication and focus to achieve his goals in life.

CHAPTER 1

Future is Beautiful

*"Only those who risk going too far
can possibly find out how far they can go."*
– T.S. Elliot

The IT service industry has evolved into something beautiful, where buyers genuinely respect the service suppliers or partners. The *Apples* and *Googles* of the world are scurrying for manpower as the service firms are calling the shots. Both the margins and revenue productivity of service firms are at an all-time high, with attrition as perennially low. The IT buyers are extremely happy as the service partners are helping in reducing the overall cost of ownership for the IT operations. Every penny in cost benefit is visible to the business buyers; and hence, it can be passed on to the end customer to retain old and attract new.

Am I daydreaming? No; that's the reality, and the world is so beautiful for everybody. Before the critics are out analysing the feasibility, let's enjoy the moment. If we cannot dream big or different, it will never be a reality. There will be a few of my reader friends who would be contemplating, "Is there a problem with the present?" No; however, if it's good now, it can be better. So let's talk about some of the observations that make the future beautiful.

Let's start with our remit of service:

1. We are providing service; yes, you read it right!
2. We are helping firms in delivering modules, or functionality, or pieces of software.
3. We are known for our capability, not for the number of resources we have.

We are seen in the industry:

1. By the number of successful deliveries we have done.
2. By the number of outcome-based implementations we have achieved.
3. By the number of innovations we have brought in:
    a. individual solutions
    b. partner-led solutions
    c. transformational solutions
4. By increasing revenue productivity.
5. By the number of clients we have helped to be successful, in the true sense.
6. By the number of staff augmentation firms we have seconded with, for individual roles.

We have been seen by our buyers or clients:

1. We are present as part of their think tank for the future.
2. We are asked for advice on certain business realization of goals.

We are seen by our employees:

1. As a firm to work with, come what may.
2. As a firm with the highest intellect to challenge the younger minds.
3. As a firm they could associate to fulfill their desire for excellence in IT
4. As a firm they are proud of!!

The buyers have evolved:

1. The IT buyer has the best representation of consulting, delivery, and operations from the key partner.
2. The IT buyer is assessed as best as per any maturity model in the software industry, if they are leading the partners(suppliers).
3. The sourcing buyer has a robust framework to assess the IT buyer, as well as the suppliers (partners) on delivery.
4. There is a clear strategy on what to keep in-house and what to out-source, so there is sufficient knowledge or know-how present in-house to regain control as and when necessary.
5. The IT buyer has enough know-how to augment the partner work force in case of adversity.
6. There is one standard or industry standard applied and practiced so that any delivery from partners can be assessed objectively.
7. The business buyer is keeping all the partners (suppliers) aware of the end objective rather than allowing it to be lost in translation.

Is it too much to ask for? Yes, in current context, I feel the industry needs some guidance; of course, not from me. The answer is there. My earnest request is to look around. You will find the answers.

Before we get into the answers, let me put forward some questions. Trust me, you will find the reason why I am raising these.

Let's first talk about the suppliers; then I will pose similar questions from the buyer's perspective:

1. Does your employee love their job?
2. Are they really growing?
3. Is it only money that dictates attrition?
4. Do you have answers for those who aspire to be part of the *Googles* or *Apples* of the world?
5. If attrition is a problem, have we raised it with the buyer?
6. Can there be a consortium of service providers who can agree on what's right and what's wrong?

Now, from the buyer's perspective:

1. Do you engage with the suppliers directly?
2. If not, does the IT buyer layer have enough control in place to make decisions?
3. Does your IT buyer layer lead or follow?
4. If it's a lead, then on what basis: maturity models, organization transformational capabilities, or total cost of ownership?
5. If it's none, or only one of them, the way a consulting partner's view is taken, a service provider's view is equally necessary.
6. What is the role of the sourcing buyer?
    a. Is it only to augment the IT buyer's decision or bring in some matrix to objectivity?
    b. Do they have the independence to put forward their point of view in decision making?
    c. Do they have the framework and know-how of maturity models to put forward their assessment?

My dear friends, my objective is to put forward some key questions. Some of them, you might not like or might not agree with. But the reason I am keen on bringing it forward is because everything has to be ironed out if we are eyeing for a beautiful future.

If you remember, I mentioned about how the answers are there, out in public. Let's reveal the facts, and I am sure you will agree with me by end of this section.

Let's look at the phone manufacturing firms: OEMs and the ODMs.

OEM stands for Original Equipment Manufacturer, and ODM for Original Design Manufacturer. In the case of Apple, it's the ODM; whereas Foxconn, in China, is the OEM. Apple spends millions of dollars on R&D of the next new gazette, and then outsources the same to Foxconn to assemble parts from other prescribed suppliers or OEM partners of Apple and iOS, from the OEM supplier, Apple, itself.

Then there is Micromax, a brand in India, who is an OEM, and asks an ODM, called Qiku, to manufacture based on some key requirements of a smartphone. Qiku comes with a phone, or already has a similar phone that Micromax is asking for. Micromax asks to put the Micromax logo in the phone, in the manufactured piece of Qiku.

Lastly, TCL-Alcatel (OEM) builds smartphone handsets for multiple clients like TCL, Alcatel, and Blackberry (ODM), with Android OS from Google, or Blackberry OS.

For both IT supplier and buyer, my question remains the same: What do you want to be?

Although we have not spoken about people in any of these examples, it's the outcome or part of the product.

CHAPTER 2

You Have Been FAANGed ...

In the last two decades, I have worked as a supplier and have had wonderful opportunities to work with lots of clients—small, big, established, upcoming, market leaders, disrupters, aggressive, passive, low maintenance, high maintenance— wonderful colleagues from competitors, partners, and analysts across geography.

During my growing years in the IT service industry, Y2K loomed large, followed by Euro conversion, with a deadline of 2001 rattling Europe. With the dotcom boom throwing businesses off track, a new frontier was shaping up in the disguise of South Korea, Israel, Ireland, and India in IT. China was getting stronger with every passing moment in manufacturing, resulting in the mass liquidation of colonies of manufacturing hubs in the developed world, and increasing the load on local government due to social security.

As a prodigy, this book is my two cents. If this can help any of the firms in the world anywhere, I will be proud of myself.

As a supplier, the priorities are:

- Value for Money
- Automation and Innovation
- Predictable Delivery
- Best in Quality Output

In return, what the buyer can offer is:

- Bidding, even after agreed rate card
- Ask for *bodies*, or human resources, rather than service
- Multiple rounds of interviews before selecting human resources from supplier
- Expecting supplier or partner to deliver with resources chosen and managed by buyer

It's like when you have asked your builder to build a house; however, with the labor and material, you will decide whom and what to deploy, within a budget defined by you. I do not foresee the project being completed ever.

Anyway, I wondered why there are so many disconnects, and more so, why there is no acknowledgement or recommendation of resolution of the issue. I spent long hours on the electronic media, printing media, and search engines to find out why this is never being reported, or why nobody has ever tried to address the ambiguity.

So if you feel that there is clarity on this at every level in your organization, then you are probably part of a mature organization. But as the problem is so deep-rooted, it involves organization level awareness to solve the issue.

It's all the more ironic that even after 50 years of existence, IT services is often misunderstood and trivialized to *staff augmentation*. This raises two sets of fundamental questions—one for the buyer and one for the supplier. The questions for the supplier are more on intent, as we will cover the "how" part in due course.

Questions for the Buyer

- Are you buying capability or service?
- Is this the *right* supplier?
- Does business choose IT staff, or expect a certain outcome from CIOs?
- Are we leveraging the supplier/partner in the right way?
- Do I know how my sourcing strategy is being implemented on the ground on a daily basis?

- Are we listening to our supplier(s) vis-à-vis selective hearing?
- Does the supplier(s) have a platform to express or advise us?
- What's in it for the supplier?
- Does the IT buyer need specific training on supplier or partner management as sourcing buyer?
- Do I need a partner? If so, am I ready to bring in change in my organization?

Questions for the Supplier

- Are you ready to be a partner?
- Are you hedging too much on buyers, though *strategic* on paper but *transactional* in reality?
- Do you see an intent from your buyer to move into a partnership?
- Are you invited to forums, or privy to any strategic discussion where business, IT, and sourcing buyers participate?
- Are your goals aligned with the buyer's strategy?
- What's your strategy for getting out of the *sea of sameness*?

Sooner than later, the IT suppliers would need a success story like Apple or Google. Otherwise, before they realize it, an ideology similar to Amazon could wipe out the way it's panning out in the retail industry.

Partnership is almost inconceivable without the intent and consent of the buyer. So, in my book, I have addressed the *buyer* as the *senior partner* in the relationship.

Never has there been, in the history of technology, a period as exciting as now. We are in a phase of evolution where, with every passing day, the importance of the Industrial Revolution, back in the nineteenth century, is diminishing, with the new revolution of *Artificial Smartness* being introduced to the machines. The norm is being challenged, and new business ideas are born and executed at regular intervals. With the advent of web, software, and digital, the customer centricity is redefined in the new millennium. The pace of change is so fast; either you keep up with it, or the writing is on the wall for your business model.

The threat is no more limited from a set of competitors of the same business line; it can come from the most unexpected player. There are countless causalities since the year 2000: when TomTom was *Googled*, Blockbuster Video was *Netflixed*, the payment industry was *Appled*, and retailers were *Amazoned*. Firms like Nokia, Kodak, and ToysRus—pioneers in their respective business once upon a time—were decimated by technology firms rather than fellow retailers or telcos.

So, the most quintessential learning from the recent happenings in the market place is to *"be relevant, or soon you will be obsolete."* A bank or retailer should worry about how to protect the customer base, or improve the customer centricity: Be a disruptor on your strength and business offerings, and/or adopt disruptions in technology fast. There is no shame in taking advantage of disruptions that have already happened, rather than re-inventing the wheel altogether yourself. If Facebook, YouTube, Google, and Uber can be the market leaders without owning any significant assets, why can't you!!

Before your business is *FAANGed*—or in other words, your business is eaten for lunch by the likes of Facebook, Apple, Amazon, Netflix, and Google—an immediate course correction is imminent.

Immediate Problem at Hand – Bullwhip Effect

The *bullwhip effect* is the uncertainty due to distorted or no information flow, up and down the supply chain. If I have to borrow this in the IT service industry, then let's look at the parlance.

What causes the bullwhip effect?

- Break in or no communication through supply chain
- Poor or no forecasting of demand
- Price variation midway

What are the implications for the supplier, as well as for the distributor/producer?

- Excess inventory

- Increase in cost base for supplier
- Impact on quality
- Lengthened lead time for future
- Unnecessary adjusted capacity

It's clear from the implications that the impact is more pronounced in ascending order, from retailer/distributor, manufacturer, and supplier. However, the overall impact of the effect can bring in serious adversity for both the retailer and the manufacturer, if it starts recurring more than once. In this example, the manufacturer is the internal IT team and the borrowed supplier teams.

It's not entirely paradoxical to claim that all my dear readers understand the buyer and supplier roles, and what to expect from each other. However, there are proponents of the relationship that might not be so obvious. So I hope you would be looking forward to the book and the next chapter, where we will slowly unravel the origin and main actors of the journey.

CHAPTER 3

The Conundrum – Service Provider

Fast track IT journey: As the name suggests, I will be discussing the journey of IT vendors in IT services, or IT consulting services space, from the point of sale to establishing a long-term partnership with the buyer.

Before we delve into the details, let's take a look at the 3 most relevant questions, and what the answers are.

- What is being delivered? *Service.*
- Who are the key players? *Consumers, buyers, and sellers.*
- How can it be delivered? *Through partnership or collaboration between the buyer and seller.*

What is a service, and how is it different from any other means of selling? Service is a repeatable experience provided on a continuous basis or on-demand to the consumers or customers. The user experience for a service is an amalgamation of products and services, packaged to deliver a defined set of utilities for user convenience. What are the different types of services, and who are the service providers?

Hospitality – It's a service provided through infrastructure or real estate, like hotels, restaurants, or pubs, or by transportation means, like trains, cars, flights, or cruises. The consumables are food, drinks, or comfort, supplied by a whole host of suppliers and providers.

Be it TV or Wi-Fi, Android or Apple, gas or electricity, the service is provided as a package through a group of people working tirelessly in the provider's organization to ensure service continuity, quality, and economy.

Healthcare, manufacturing, automobile, or construction – These industries provide both product and service, and sometimes provide experienced people as part of the service. Though the roles of the people involved are important, one cannot ask for a group of builders, designers, or painters as a service only.

That's where the similarities of other services, with IT services, end.

Let's look at the Genesis of IT services closely. Since the beginning of the 19th century, with the up rise of the Industrial Revolution, there was a necessity of advisory for every trade. This is a group of subject matter experts in a particular area, which started guiding various businesses in day-to-day management and innovations, as applicable. That's how the consulting services started, and with the ongoing success, became a popular profession. This has grown into a mature professional service, with a turnover of approximately 300 billion USD, as of 2017. Technology Consulting, or IT Consulting, is one of the key industries. There are various offerings as part of the idea of consulting. Though, during the inception days, there was clear demarcation between the companies providing non repeatable services (IT consulting), and the firms providing repeatable services (IT services), in the current market, the dividing line is blurred.

So, in the current context, IT services offers repeatable guidance on strategy, architecture, implementation, system integration, data analytics, security, and both development and maintenance of software and infrastructure.

As per Gartner, *"IT services refers to the application of business and technical expertise to enable organizations in creation, management, and optimization of, or access to, information and business processes."* In other words, IT services enables businesses to solve a business problem through the application of digital or technology. IT services started in the 60s; since then, it has grown exponentially, both in breadth and depth.

In 1911, in the US, a new company, Computing-Tabulating-Recording Company (CTR), was launched as a merger of 4 different but relevant ideas patented in the late 1880s. The firm was re-branded as International Business Machines, in 1924. Truly, IBM is the first *seller* or *vendor* that introduced machines that can solve business problems.

In the 1960s, a new type of offering became extremely popular. A group of subject matter experts, or consultants, helped the business or corporation for a fee—that's how consultancy gained popularity as a new seller or provider in the IT services arena. This sector evolved into management, HR, finance, IT, and business.

In the last hundred years, IT services has diversified into multiple streams, like technology consulting, implementation, integrators, and support.

There is a genuine problem in understanding the difference between a *service provider* or a *people provider,* in the IT Services industry. Often, the buyers (not the consumers) get mixed up; hence, the role of a service provider is diluted to nothing but a recruitment agency with preferential rates for expertise.

My fellow readers, if you are a supplier and are providing skills or resources only for staff augmentation, or have sold a product, then you are already a sourcing partner or product partner. The issue taken up in this book is for the remaining suppliers who have been chosen as service providers, and the buyers who have selected the suppliers. I believe the definition of service has to be looked at again, by both the buyer and the supplier. The example below provides further clarification of the argument.

In the next two sections, we will talk about the key players and the different types of players in the marketplace. I am sure you will like the subtle differences in their roles and expectations.

Buyer – Consumer or Customer

> "It is not the employer who pays the wages. Employers only handle the money. It is the customer who pays the wages." – Henry Ford

In the context of this book, an IT buyer is a professional purchaser or a group of purchasers, who invites propositions or proposals, and selects one or more from the group of potential suppliers, vendors, or service providers, to deliver utility, product, software, hardware, or services with the best of quality, and within budget.

These professional purchasers can be of 3 types:

1. Representative of the business unit who provides service to the end customer or consumer
2. Representative of the IT function within a business unit, who mediates or buys on behalf of the business unit
3. Representative of the sourcing function that runs the procurement process for the IT function, who buys on behalf of the business unit

Though the beneficiary is the end customer or consumer, the key decision makers in terms of influence and authority are the business buyer, the IT buyer, and the sourcing buyer. Each buyer might have separate KPI or mandate, aligned to the vision or strategy of the buyer firm. Hence, the procurement process in IT is quite complex and cumbersome compared to other industries like retail or manufacturing.

The business buyer is driven by the business strategy, which is manifested around improving the brand value and allowing the firm to keep their nose ahead of the competition. Hence, the customer base is protected, new customers are added as planned, and more so, the shareholders are happy if it's listed. As long as the venture is profitable in line with guidance, the business buyer is upbeat on investment in IT function.

The IT buyer follows the IT strategy. The potency of the IT strategy revolves around enabling the business buyer, with the best in class IT service, in a timely manner so that the competitive edge is maintained. To ensure the profitability is intact, the investment in IT should be smart and within budget allocated. Innovation and automation are the hygiene factors to improve the revenue productivity. The IT buyer also carries additional KPIs to keep the mill running in adverse events, which may be economic, political, or regulatory sanctions.

Barring some small organizations, the rest all have the sourcing function, which is constituted by a group of sourcing buyers who facilitate the procurement process on behalf of the business buyer and the IT buyer. Additionally, they are accountable for the sourcing strategy.

The sourcing strategy has 3 goals. The primary goal is to devise a plan to fulfill the additional capacity required for the IT buyer, in a duration, to deliver the service to the business buyer. The second goal is to run the procurement process for additional capacity, software, and hardware, and negotiate with potential suppliers for the best deal at the right price. And in some mature organizations, the third goal is to ensure that the quality of delivery from suppliers is maintained and improved during the tenure of the relationship with each supplier.

Seller – Vendor or Supplier

"Suppliers, and especially manufacturers, have market power because they have information about a product or a service that the customer does not, and cannot have, and does not need if he can trust the brand. This explains the profitability of brands."
– Peter Drucker

A vendor or supplier, in IT services context, is a provider of either utilities, products, hardware, software, or services. The relationship starts with a formal bidding process, run by supply chain management or the buyer; and one or more suppliers are selected to provide service for a particular duration within the mutually agreed level of quality.

For a provider of products or hardware, the relationship is more transactional in nature, and often stays warm till the deployment.

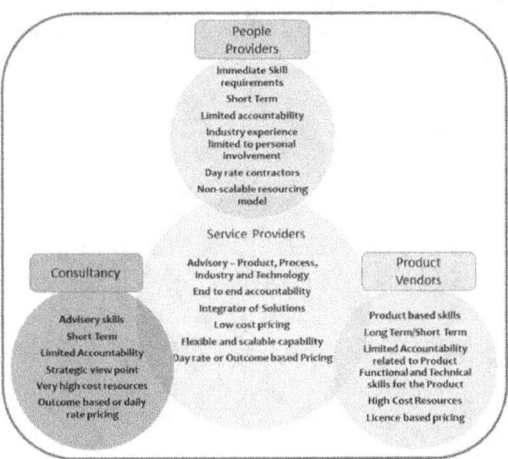

Once deployed, then the role of a service provider kicks in; firstly to run or maintain the deployed hardware or software, and secondly, to build or enhance in the future, based on the buyer's need. They are also called *system integrators*, who supply additional capability and expertise in terms of IT professionals, to manage infrastructure and applications landscape. This engagement or relationship starts through a selection or bidding process; however, the contracted duration is mostly 3 years to 10 years, depending on the buyer organization's sourcing strategy.

The final category of supplier is the utility provider, who comes up with a software solution that addresses one or more defined business problems through one, or a group of, software tools working together. In pre-millennia, the system integrators used to provide this service; however, of late, a flurry of FinTech and RegTech firms came into existence to fulfil the demands. They are also called *disruptors*, and the offerings are new age, based on the latest technology and, quite often, nimble in deployment and future enhancement. The initiation of this type of relationship with the buyer can be through bidding or through *proof of concepts*, run by the system integrators; hence, short-term in nature.

Based on the nature of service provided by different suppliers, and the duration of engagements, I will be discussing the system integrator's journey, from being a vendor to partner, during the tenure of contract with the buyer. I am sure you will be eagerly looking forward to the next chapter, where we will dig deep into the formation challenges, roles of buyers, and above all, the 7 stepping stones to achieve *nirvana*: partnership.

CHAPTER

Transformation Journey

"Transformation is a process, and as life happens, there are tons of ups and downs. It's a journey of discovery; there are moments on mountaintops, and moments in deep valleys of despair."
— Rick Warren

In 1993, J.F. Cali introduced the concept of supplier partnership, for the first time.

A mutual, ongoing relationship between a buying firm and a supplying firm involving a commitment over an extended time period, and entailing a sharing of information as well as sharing of the risks and rewards of the relationship.

It's the automobile industry that started this practice first, and later, FMCG, or the retail industry, followed suit. But the adoption of this practice in the IT service industry has a long way to go. Unlike other industries, here we have the non-tangible commodity, software and knowledge resources, which may be the reason the journey takes time—and it's a complex one.

Let's look at the complexity in the relationship of this buyer/supplier equation:

- The relationship is one to many for the same service.
- There are 3 types of buyers, with *different expectations* from the relationship.
- There is no tangible product involved.

- It's between two groups of knowledge resources, from two different organizations.
- The service quality is dependent on time, capital investment, the skill set of Human Resources, and the effectiveness of the process being followed.

The relationship starts:

- As part of the sourcing strategy, the sourcing buyer selects one or more suppliers to fill the skill gap – qualitative and quantitative.
- IT buyer selects one or more suppliers to develop a new software, which will help the business buyer achieve the business goal oriented towards the consumer.
- IT buyer selects one or more suppliers to maintain the IT landscape.
- IT buyer selects one or more suppliers to reduce the total cost of ownership of the IT landscape.

As new teams are formed across buyer and supplier, the engagements can be one of the below:

- Staff Augmentation – Filling up certain skills from the supplier, and putting them in a joint team, including the resources from the buyer, single supplier, or multi suppliers
- Managed Service – Carving out a work package to be delivered by the supplier

- The contract between the buyer and supplier, for both types of engagements, can be based on different sourcing strategies:

- Partially Sourced – As part of this strategy, the teams are hybrid.
  - Supplier provides the capability resources. Buyer keeps the control or management roles.
  - Supplier provides the control roles, and either buyer or second supplier provides the capability roles.
- End to End – Supplier fills up all the necessary roles and skills to deliver a defined service. The team is formed with supplier resources only.

So the success of the IT service is heavily dependent on teams. It's quite apparent from the explanations above.

Every relationship starts with two teams interacting, but with time, more teams are formed on both sides to deliver different facets of services. So, let's look at the 4 key stages of teaming, or group development. This applies to each team formation.

The first stage is called *Forming*—when the group members from buyer and supplier(s) are getting to know each other. In this stage, the members get familiarized with the strength and role of each member. It's more like a briefing session before a flight, on the ground. The pilots, crew members, and flight engineers get to know each other, as well as the details on destination and type of aircraft. Subsequently, on the aircraft, the flight maintenance engineer, catering staff, and airline personnel join the pilots and crew members, and share other details, like weight of luggage loaded, quantity of fuel filled, and a generic health report of the aircraft. So mostly, in this phase, the data shared is static or pre-decided.

The next stage is quite interactive; hence, the most difficult one. It's called *Storming*. In this phase, the team starts sharing their point of views and opinions on how to make it a success. When intellectuals share their view points, there will be conflicts and differences in opinions. Some will fall out, some will continue, and some will follow others. At the end of the day, you have the clearly defined leader, the thinker, and the executors in the team.

That takes us to the stage where everybody in the team is aligned with their roles and responsibility, and the focus shifts to the execution. It's called *Norming*. It's the best phase, as the vibes will be quite positive. The contrary is also a possibility, so the team is completely dismantled. An analogy can be drawn from a cock-pit drill here. The pilots and crew members go through the drill before take-off. Till the team is happy with each and every aspect, which can affect the performance of the aircraft in the air, the flight will not take off.

"When you start out in a team, you have to get the teamwork going, and then you get something back."
– Michael Schumacher

Please fasten your seat belt, as we are heading to the last stage, where the team is about to take off as it starts performing. Though it sounds like we have reached the pinnacle, I'm sorry to disappoint you. We have just completed the first step in the buyer/ supplier relationship. When two organizations are on a joint venture, the bond is quite weak till the first success is accomplished and celebrated. Each success adds to the confidence in the partnership. Is continuous success the recipe for a buyer/supplier relationship to transform into a partnership? The short answer is, "Yes," but the long answer to the question is, "Sustaining success on various parameters over a period of time is the key."

The devil is in the details. When multiple buyers, with different strategies and goals, liaise with one or more types of suppliers or partnerships, with different strengths, there are numerous touch points. Please visit my book website to download the details on Multi-dimensional Challenge, and the ways to detangle the problem at hand.

Now, coming back to the formation of the team, it takes two to tango, right? We have a scenario of the single buyer, with anchors of multiple strategies, and then we have the supplier. Both have a role to play in this journey. But the buyer is the dominant partner, and the supplier is secondary. So let's talk about one of the inspirational stories from Steve Jobs.

Liquid and Grit

When Steve Jobs was young, he met with an octogenarian—a widowed man, living up the street. Steve was helping the old man with chores, like lawn mowing. One day, he called him to the garage and showed him a dusty old rock tumbler created with a coffee can and a band attached to a small electric motor. The old man then took out a few small stones from the garden, and he put them in the can with some liquid and a bit of grit. Then he switched the motor on and asked Steve to come back the next day.

The following day, Steve arrived, and they both went to the garage. The old man stopped the motor and opened the can. To Steve's amazement, the same common stones, rubbing against each other, transformed into beautiful and polished stones.

This is being used in many talks, where it shows when two or more talented people brush shoulders while working, the outcome can be quite spectacular. So, bringing in the same metaphor here, when two groups of incredibly talented people are left to rub shoulders with each other, the individual ideas are polished into collective success.

Taking the leaf out of the story above, we will go through the "liquid and grit" necessary for the buyer and supplier to take the relationship to a partnership.

In the beginning of the chapter, we had reiterated the importance of the buyer in the transformation journey. The buyer, being the dominant partner in the relationship, has the most say in the plausibility of the transformation.

Firstly, does the buyer intend to convert the supplier relationship to a partnership?

If the answer is *yes*, then there is a case worth pursuing. If there is an intent, then that takes us to the "How" and "What" aspect of the answers.

So let's look at the ways in which the buyer can help or facilitate a supplier's journey to partnership:

- Sharing – Information on strategy and goals shared with the supplier will help both the parties in proactive planning of capability building and knowledge curation; hence, the ambiguity and spend will be minimized during execution.
- Engage – Dealing with knowledge resources, the buyer's participation and respecting the supplier resources creates a high performance team, as well as helping to sustain it for long period of time. Time invested in the relationship, both qualitative and quantitative, reflects the intent towards making it a success. Although the power balance is in favor of the buyer, intellectually it's a relationship of equals. Maintaining at every level in the organization will help the bond grow stronger.
- Assessment – Assessment of performance, and faster feedback, hastens the rate of success.

Now let's take a shot at the supplier—the provider in this relationship. There are 7 aspects in the journey, from the supplier's perspective, which needs attention.

It is time to introduce the **7C Principle**, for a supplier organization to practice, improve, and sustain the momentum.

- Customer Centricity – Understanding the buyer's strategy, helping the buyer achieve their short/medium/long goals through joint planning and supplier offerings.

- Communication – Effective, creative, transparent, and continuous communication is essential for establishing one's credibility in partnership

- Creativity – Dosage of creativity adds new sparks in a relationship. It's the catalyst in the transformation journey.

- Commitment – Seamless accountability across partnership; suppliers intention to "put the skin in the game."

- Capability – Sustain high performance, de-skill and re-purpose faster and efficiently; periodic upliftment to maintain relevancy.

- Collaboration – Going the extra mile, within all ethical and legal boundaries, to bring collaborative offerings that are beneficial to both.

- Continuity – Push to better the best; shorter feedback cycle; continuous review.

Don't go away, as we will delve into details of the above aspects, one by one, in the next few chapters. We will also look at various case studies and events in the business world, which have transformed business relationships into sustained long-term partnerships.

CHAPTER 5

Customer Centricity

"Everyone says Toyota is the best company in the world, but the customer doesn't care about the world. They care if we are the best in town, or not. That's what I want to be."
— Akio Toyoda

In the current context, the origin of *market*, or *trade*, is dated back to 3000 B.C., when Cyprus Island became a prosperous place on earth. The exotic deposits, found on the rocks, were named and shipped as *Copper*. Later, around 1000 BC, another precious metal, called *Tin*, was found to announce the beginning of the Bronze Age; and Cornwall became one of the first key European suppliers of the millennium. Even the inception and success of Silk Road, in the 2nd century B.C., is a testimonial of the importance of customer centricity.

So, what's the big deal about the term, *customer centricity*?

Customer centricity is the equivalent of the solar system or the lunar calendar, where the customer is at the center of business. The actions, ideas, and executions all depend on the need, necessity, and passion of the consumer or customer. Both the success and the extent of success are at the mercy of the end customer. Is it? Or does it depends on the "What, How, and Why" aspect of the solution, which is offered to the customer.

"What" addresses the problem statement, whereas "How" creates the customer experience, and lastly, "Why" establishes the customer loyalty.

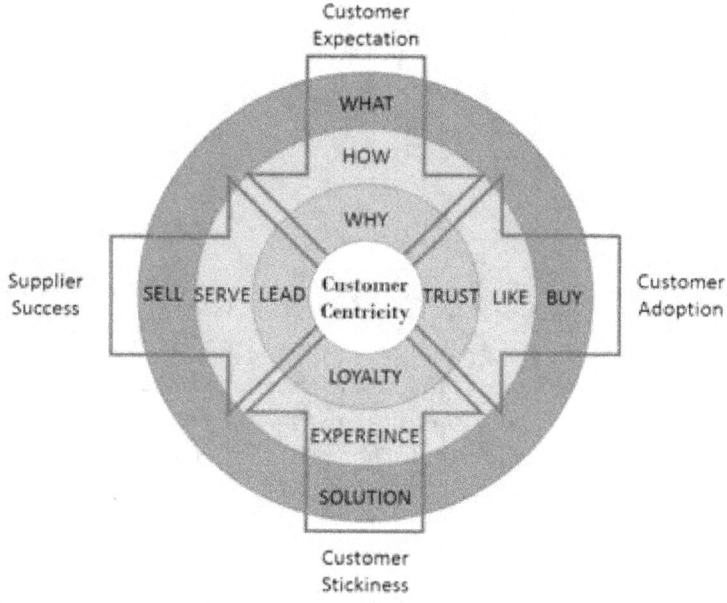

The Golden Circle, as explained by Simon Sinek, is the decoded version of customer or consumer loyalty. Whenever a product or service firm comes up with a new idea or product, it goes through the basic "What, How, and Why" framework. The answer to the "Why" or the inner circle, defines the extent of success. Extending the same argument on customer adoption and stickiness, when the consumer or customer trusts a brand or is being loyal, then the features (What) become less relevant. It's more of a subjective choice of how you feel, and whether you believe in the brand or supplier.

Bringing the parlance here, for a supplier's success in business and acceptance towards partnership, the answer to the questions are vital. Does the buyer see and understand the "How and Why" in you? "What" is expected anyway, so the facts and figures can prove your worthiness; however, the maturity and longevity of the relationship is always being decided by the subjective parameters.

Let's start with the first subjective parameter—customer loyalty—which is the offshoot of when the buyer starts believing in the "Why" of the supplier/seller. There are hundreds of parameters out in the market to analyze and improve customer loyalty. But the key remains in the subconscious part of the brain, which pushes us as the buyer to go with our instinct for a particular brand or a specific name. So, either the buyer wants to be associated with this seller's brand, or the buyer's gut feeling tells him that there will be nothing wrong if they choose the brand. It takes time, but the seller or supplier has to be obsessed to establish and continuously improve on its brand value so that, over a period of time, the buyers would love to associate themselves with the seller's brand; hence, the beginning of a true partnership.

> *"You should not build your customer service system on the premise that your organisation will never question the whims of your clients."*
> *— Richard Branson*

The second subjective parameter — common objective: This addresses the "How" aspect. I have restricted my explanation below for service providers only.

In a partnership, both the buyer and the seller need to understand each other's objective, or the short-term, medium-term, and long-term goal for a particular time frame. So let's talk about the buyer's objective first.

- Understanding the buyer's strategy
    - Business strategy
    - IT strategy
        - Best in quality
        - Advanced/cutting edge
        - On time delivery
        - Within budget
        - Lowering TCO
    - Personal growth strategy
    - Sourcing strategy
    - Capability strategy
- Understanding the buyer's short/medium/long-term goals
- Mapping the buyer's strategy to vendor strength
- Joint planning
- Agreeing on clear outcomes

Most organizations will have a well-defined, time-bound vision, supported by a well- thought through strategy. The execution of the strategy will be realized through two ways. The tactical way of realizing the strategy is to come up with short and practical goals that can be achieved in a shorter time frame of 6 months to 12 months. The more complex and transformational goals will have a longer time frame of 2 to 5 years.

Most of the time, the visions of competing buyers are similar in nature. Whether the buyer belongs to the manufacturing industry or financial industry, the most common vision is to be either the best in the industry or the most respected brand around. If the buyer is a listed public company, one of the common visions is to have a lower cost to income ratio; alternatively, the earning per share should be higher.

A vision or mission, without goals and objectives, is like a boat sailing in high winds, with a broken or nonexistent mast. To achieve the vision, every organization or individual will often come up with a list of goals or objectives, which are measurable. Both quantitative and qualitative tracking of the goals helps in assessing the plausibility of the organization's accomplishment towards the vision.

Now, to realize the vision, the buyer comes up with different strategies (e.g., business strategy, IT strategy, personal growth strategy, sourcing strategy, and capability strategy). While business strategy concentrates on increasing the customer base and the revenue, the IT strategy focuses primarily on creating the best in quality software platform, using the most advanced or cutting edge technology, in a defined timeline, within budget and with minimal cost of maintenance.

The next three strategies revolve around one of the key stakeholders: people. Every buyer organization will have a sourcing strategy that mandates the optimum split— between capabilities to be hired as permanent employees, and the capabilities that can be fulfilled by partner organizations or suppliers. The second strategy involves human resources and capability strategy. As part of capability strategy, organizations ensure that both the core capability and the flexible capability provided by the partners are enabled enough to deliver the business and its goals, laid out as part of the other two strategies.

One of the key strategies, which is mostly overlooked, is the personal growth strategy of the key executives who are directly accountable for the fulfilment of the visions. This could be a maverick CIO, an ambitious business director, or a charismatic CFO of the buyer organization.

Strategy is the method, but a plan to execute the strategy is a list of actions to be completed to achieve the goal as per the vision. When a relationship or team is formed, to ensure the longevity and effectiveness, there has to be similarities in goals, or a common objective. Alternatively, the actions on both sides should be complementary, to improve the productivity, or to enhance the outcome.

> *"Every once in a while, a new technology, an old problem, and a big idea turn into an innovation."*
> **– Dean Kamen**

The third aspect in customer centricity is around the question of "What", so it's about facts and figures. Here, the supplier should concentrate on one big idea and pursue it like there is no tomorrow, so that the business value realization is quite clearly visible to the buyer. Every promise made towards successful delivery of programs, initiative, or big ideas, should be tracked and reported periodically to ensure the trust is built and maintained, day in and day out.

Let's look at one of the success stories, on how objectives can be common or complementary.

Amazon – Customer adoption holds the key for an innovation to mature as a disruption!!

Amazon has been around for 22 years now, and it's a story of transformation from a small startup to a giant. Amazon has transcended into history books as one of the amazing success stories of the new millennium.

What went right for Amazon, or rather, what did Amazon do differently?

Vision – Customer Obsessed

4 key objectives

- Customer obsessed – Here, *customer* means the *end customer*. Keep an eye and ear on the end customer, and then lead or help both business and IT buyers to hear the end customer. The key is *what the customer wants*, not what the business house or their IT division thinks is right or wrong.
- Invent and Pioneer – Jeff Bezos strongly believes in disruption, but in a slightly different way. Innovations or inventions are not disruptive; it's the customer adoption that makes it path-breaking in a true sense.
- Long Term Play – Amazon moved away from assessing or planning for a Quarter; rather, always looked at the long term, like 3 years, before writing off a strategy. Every firm should believe in their strategy for 3 years or more.
- Continued Success – Fear or failure may lead you to mediocrity. *Fail fast and scale stable* is the mantra to use in order to see continued success.

3 key messages

- Big Idea – Rather than spreading bets in multiple small ideas, pick one or two big bets; however, live/breath/die by the big bets, to make it a success. Move away from the also-ran mentality with hedging all the investment in mediocrity.
- Enforce the Obvious – Don't lose focus on the basics. Don't let the big idea overwhelm to miss out on common sense, like value for money, fast or on-time delivery, de-selection on nonstrategic services, and predictability of delivery service for a long time.
- Sustainability – Changing the supplier and seller behavior to use lesser and recyclable packaging, so that Amazon has achieved in reducing the wastage by around 55 thousand tones for year. This demonstrates empathy for the environment as well as creating newer ways to optimize the usage of packaging material, which ends up as trash all the time.

While the retail market leaders were busy with competitor market, business mobilization, product, or technology, Amazon was betting on customer obsession. Even in technology, before anybody could see the benefit of a newer idea called *Cloud,* Amazon jumped into it and became a market leader in *public cloud*.

Any idea or initiative, if not communicated, or not communicated effectively to the right recipients, would be difficult to plan or implement. That takes us to the next chapter, where we will discuss the importance of the next "C": Communication. I will also be sharing an interesting invention of mankind, which changed the ecological balance forever on earth.

My 3 Cents

1. Create your "Why/How/What." How does it map to each customer's objective? Create your annual plan, and try to be as objective you can. There should be a number of actions, which should help you in assessing the accomplishments and learnings to be used for subsequent year's planning.
2. What are your big bets? How are you faring on the big bets? Do an honest assessment every year, apart from periodic progress tracking.
3. The next action is the most important: Share your "Why/How/What" plan with your customer, pre-post, as that would help you in aligning your objective with the customer.

CHAPTER 6

Communication

"Effective communication is 20% what you know and 80% how you feel about what you know."
— Jim Rohn

Where the world is the marketplace, the discussion is that both parties in the world should make an effort to understand each other, and to get the most out of every interaction. In an international business context, with people on both sides, coming from two different backgrounds or two different objectives, the words spoken, the emotions, and the lack of understanding of each other's cultural values will soon become detrimental.

Let me share a few examples to elaborate the sensitivity in a cross-cultural scenario. In Asian culture, being hierarchical, a question or a challenge is perceived as a negative attitude. Whereas, in western culture, a question or a challenge is often used to either clarify or to seek an alternate view point.

In the global market place, fellow Americans could be seen as loud, whereas Canadian cousins would be still struggling to get a response from the English on their humor. In Japan, China, and India, grey hair or hierarchy commands respect, whereas an Aussie's supplier might be too hot to handle for an African buyer.

The diversity is so pronounced, yet it's a wonderful world, full of intelligent souls who are engaged in successful business ties. Whereas there are unwritten codes of conduct, passed from generation to generation, on what to say and what not to, Asia is proud of its history and cultural heritage in bowing to elders.

So it's quite natural that both the parties have to give each other some time, and a couple of chances to understand, before bringing out the negativity of calling each other incompetent, or passive or aggressive.

> *"To effectively communicate, we must realize that we are all different in the way we perceive the world, and use this understanding as a guide to our communication with others."*
> – Tony Robbins

Communication is the bedrock of any relationship. The success of a relationship depends on mutual trust, mutual benefit, and consistent reassurance. Communication is the most effective tool to express reassurance; hence, a confident relationship is built with time. In business, the relationship can be transactional or strategic; however, it evolves for good through effective communication.

There are numerous frameworks and tons of material available, both in electronic and print media, on communication. Hence, I will restrict my discussion to only key aspects of communication that are relevant for a supplier to alleviate to partnership. In a buyer/supplier scenario, the communication is often biased in one direction. You know, it's that way, isn't it? It's always a reactive communication from the supplier. The buyer often builds a perception towards this behaviour of supplier as Order Taker. Whereas in partnership, the communication is in equilibrium. It's a conversation between equals. Be direct and truthful, but not rude, of course!! No preambles to justify your answer. In a global market place, the cultural difference plays a vital role in how we communicate. Hence, certain exceptions apply in messaging, not on the content of the message!

Communication should be continuous. The buyer/supplier conversation should be on a regular interval, at every level of hierarchy. It can be in person, over mail, through meetings, or in a social or corporate forum. This way, the communication is always in present and forward looking. On top of this, if there is transparency in the interactions, it negates the possibility of any adverse perception seeping into the partnership. That brings us to the next important aspect: how to engage.

> *"The art of communication is the language of leadership."*
> – James Hume

TIE – All our conversations are one of these: *transactional, intellectual, or emotional*. The communication between two organizations is led by human beings; hence, the framework is quintessential. Transactional communication is the junior order of communication, where certain data is transferred, and the outcome is often binary in nature, either *yes* or *no*, or sometimes, *do not know*. It's for a particular transaction, and is short-lived. It does not create any long-term impact on the relationship. The other two types are more engaging, as these are related to the human brain. So, connection at all three levels makes a communication engaging, and both supplier and buyer are receptive to each other.

That's the "Eureka" moment!

We are all blessed with the power of our brain—from both sides. So let's leverage our right brain and be creative. Believe it or not, we all have this trait. Advisors, lawyers, motivational speakers, and all leaders exercise a certain level of creativity while delivering a message. It influences all, most of the times. So, how can we bring that into our day-to-day practice as a supplier? This can be achieved at two levels. The corporate level creative communication is targeted towards *brand building*. The marketing and public relation function of the organization is mostly responsible for this type of creative communication.

The second one is at the supplier personnel level. There are various possibilities. Some of examples are:

- Presentation of quarterly report to the buyer, with key highlights on how the supplier organizations fared and will do in future. What are the strategies for the next three quarters? How are those strategies aligned with the buyer's goals?
- Organizing workshops to decode some of the key regulatory requirements
- Be a catalyst to drive your customer towards greater goals

> *"Humans think in stories, and we try to make sense of the world by telling stories."*
> **– Yuval Noah Harari**

Storytelling is about sharing experiences, and is one of our oldest art forms. This art form started a zillion years ago. One of my favorite books, *Sapiens*, clearly highlights the ascent of Homo sapiens with this ancient tool of communication. It stimulates listener's imagination as words come alive with the narratives. It invokes emotions or reaction; hence, it builds a strong bond between the speaker and the listener, even if for a finite duration. But the story lingers in the psyche of the listener, long after the conversation ends.

Biggest invention after fire, which transformed humankind – Sapiens, by Yuval Noah Harari

The story goes back to 200,000 BC. There were so many different kinds of human races (e.g., Homo Denisovans, Homo Soloensis, Homo Neanderthals). Soon, the world of human colonies realized the existence of another kind, called Homo sapiens, who originated from Eastern Africa. In the next 130,000 years, or by 70,000 BC, all other races were extinct, except Homo sapiens, who are ruling the earth till date. What was the secret behind the success of Sapiens? Evolution had been working in the background on the optimum physiological characteristics of a human being who could survive or rule. Apart from their build or size, there was an invention by the Sapiens that made them invincible. In reality, the factor would sound trivial; however, in terms of the impact, the invention was enormous and earth shattering—literally!!

All right! Before we discuss the invention, let's look closely at some of the key animal instincts that we all carry, being evolved over a period of time. When a tiger, a lion, or a snake prowls on a group of animals or birds, if baboons, monkeys, or birds are around, they would create a loud noise to alert the group, so that the animals could take pre-cautionary steps to survive. Secondly, when a predator is hungry, it will go for a kill. Once its tummy is full, it will not even bother, even if the prey jumps onto them. So there was a set of unwritten rules that used to maintain the ecological balance. These rules were tweaked a bit by all the human races, with the invention of fire, and when they learned how to leverage agriculture as an alternative to the hunter-gatherer lifestyle.

Now, coming back to our original story, as soon as Sapiens started spreading across the world and got in touch with other human races, all other races were decimated. The next biggest invention by Sapiens was gossip, or storytelling, which could be fictitious, full of fantasy, and larger than life. It was a new way of creating fictional stories and spreading them among other human races to control their mind. Sapiens used this tool to create negativity through fiction or lies, and to conquer. On the other hand, they were able to use this tool in a positive way to create co-operation or collaboration amongst a group of people for a common cause. That was the beginning of legends, or history. Thirdly, it helped in the origination of ideological diversities, like religion, politics, and economy.

Storytelling is effective, as it first engages, then unconsciously persuades, and finally, influences the listener to take an action. Every great fiction starts with an adversity, which the listener or the reader can relate to. Subsequently, the gorge between good and bad is amplified. Finally, the protagonist, in spite of all his or her selfless vulnerability, finds a solution, and then the celebration follows. The solution could be a creative idea from the story teller; however, it's delivered as if it has come out of the shared experience of a group of mere mortals, or the protagonist. That makes the story more believable.

Let's talk about the contribution of the other half of the brain, the *left brain*, which is dominant with data and analytical ability. The great leader in IT, and the Infosys founder NRN, is known for many motivational quotes. The one that is relevant and most notable here is, "We all believe in God; the rest, kindly come with data." It emphasizes the importance of data in knowledge industry. In presentations, or in meetings or in mails, this form can be leveraged along with facts, data, and a solution relevant to the buyer's goals.

Tour Guide Philosophy – Think Smart and Talk Fast

To be a successful Tour Guide, what do you need? This is an interesting observation, and all of us can learn and leverage from this story. There is a unique pattern and niche methodology being followed by the tour guide to lure the tourists.

Firstly, it's the *approach* – Speak what the tourist wants to know, rather than what the guide wants to say.

The second aspect is to *know your audience*. What's the profile of the tourists—nature lovers or number geeks? Which age group do they fall under? Are there any specific areas of interest—history, spiritual, or geo-politics? Are they local or outlanders?

Third and key element is *how to make the conversation contextual for the tourists*. Whether the tour will be interesting or a drag, it all depends on the skill of the guide to deliver the message and information through relevant examples and anecdotes, to which the tourist can easily connect. If things are not going great, then the maturity of the guide can save the day, if he can react and change course based on verbal and non-verbal feedback received.

Finally, it's the *structure of communication*. There are different techniques the guide uses to enhance the interest of the tourist in receiving the information: Navigate – past, present, and future; or Persuasive – problem, solution, and benefit; or Semantic – what, so what, and now what.

Let's look at the next paragraph, which talks about creativity in communication.

Most times, we prepare for our presentations and meetings. Sometimes we bump into the key stakeholders in airports, at conferences, or on the golf course. Rather than getting into a meaningless conversation and losing the opportunity, all of us can practice some of the different and creative suggestions of motivational speakers.

What's the recipe of impromptu speech on a random topic?

- Use 3 breathless sentences to describe
- Followed by 3 repeated sentences
- To top it all, 3 balanced statements
- Drop in a metaphor to make it contextual
- With a bit of exaggeration
- End with the final say in rhyming couplets, or in a couple of small sentences

It's fun to practice; however, it is a powerful tool to improve the communication with buyers—both verbally and written. With that, we will move onto the next chapter, which details more interesting aspects of creativity. Get ready to think out of the box!! Let's go.

My 3 Cents

1. Over Communication – Is it a bane or boon? Never in business. Create a communication plan for your engagement, and ensure that both good and bad reaches the customer or client stakeholders at regular intervals.
2. Every proposition or solution – Even if the solution or technical resolution is common, try to create a nice storyline to present so that the overall story becomes your unique selling point.
3. Try to engage with the customer at all three levels: transactional, intellectual, and emotional (may be through 3 different individuals or 3 different forums).

CHAPTER 7

Creativity

"Creativity is just connecting things. When you ask creative people how they did something, they feel a little guilty because they didn't really do it; they just saw something. It seemed obvious to them after a while. That's because they were able to connect experiences they've had and synthesize new things."
— Steve Jobs

Any relationship, over a period of time, becomes monotonous, and a dosage of creativity always works as a catalyst for both the partners. Creativity is the impetus that changes the perception and expectation from each other. It invokes reaction; hence, both the partners are keen to explore the new paradigm in their relationship. It looks like we are getting romantic here. Let's get back to the supplier and buyer relationship.

Even though the relationship here is between two organizations, often it's headed or represented by two or more groups of human beings. So, every time a conversation happens or an interaction takes place with the buyer representative, the supplier can be creative. That does not mean that every meeting should be a play, and that the supplier representatives start with dramatized dialogues.

When there is a well-defined problem statement from the buyer, and a simple one, like looking for 3 profiles of senior consultants in a particular technology or line of business, in a defined location for a finite duration, starting from a particular date, every supplier will respond with 3 profiles of consultants. That's a straight fit resolution.

Is there a creative alternative? Yes, that's feasible. The profiles can be rewritten in a format in which the buyer organization is well versed. Secondly, the profiles can be reformatted with the achievements and experience details highlighted appropriately. Thirdly, a supporting mail can be sent, with a table with key details, including LinkedIn details of the individuals (with the consent of the individuals). In this way, the supplier is helping the buyer to make the decision without spending too much time and effort.

There are so many ways, with a little bit of additional effort and time, that a supplier can bring in the differentiation, even in a simple ask.

> *"Creativity is piercing the mundane to find the marvellous."*
> *– Bill Moyers*

Another example is that every engagement between supplier and buyer is contractually binding, so there will be periodic reporting. Even though there is a defined format, the buyer organization always welcomes a little deviation, as long as it's interesting. So there you go!! Please unleash your creativity and share something that the whole organization might be worried about, and you have a relevant solution, tried and tested with some other buyer. Every relationship meeting or delivery review can be leveraged to instill a bit of creativity in reporting or information sharing.

> *"Creativity is allowing yourself to make mistakes.*
> *Art is knowing which ones to keep."*
> *– Scott Adams*

With growing focus on creativity, the following are some of the aspects that we should be fostering in internal teams, and with the client.

Evolutionary vs Revolutionary – It's not pragmatic to expect revolutionary ideas all the time. However, most of modern day innovation or success stories are based on evolutionary principle. Moving from analogue to digital TV, followed by hard disk to store, subsequently applying algorithms to watch multiple programs; then finally, stop/rewind/forward live TV. The world did not stop there; it has moved further to Internet TV, and then streaming, like Netflix.

Another example is how iPhone brought in touch technology, phone, apps, and music player, all into one device. These are all *incremental evolutionary steps* over what has been achieved already by others. So, in IT services context, we try the baby steps every time we are in design phase, testing phase, or coding.

Diversity – This is one of the key ammunition at our disposal, which is often ignored. The diversity can be mostly at four levels: *experience, gender, cultural, and professional*. In any team, there should be a healthy mix of youngsters and experience. Similarly, the aesthetic sense or problem-solving trait in women is often quite unique and different.

Threat vs Opportunity – People from a different geography, location, or culture would have diametrically opposite points of view. Lastly, if the Bullet Train is inspired by extensive studies of animals by biologists, or fluid dynamics by physicists, why can the same not be replicated in software development life cycles. So there should be an earnest attempt to create teams as diverse as possible to ensure the outcome is productive and well thought through.

> **"Creativity involves breaking out of established patterns in order to look at things in a different way."**
> **– Edward de Bono**

Threat – Thinking out of the box! Literally!!

This is a short story from Italy. Long ago, there was a small business owner in an Italian town. He had a beautiful daughter. There was a money lender who was cunning and notorious. He was known for lending to businesses and trapping them in a downward spiral of never ending debt. The money lender fancied the business owner's daughter, and wanted to marry her against her will. The lender came up with a fool-proof idea to ensure as an outcome that he gets the girl, and the business owner gets rid of the outstanding loan. The lender got a bowl full of white and black pebbles. He got another bowl, which was covered, and the contents were not visible. The lender outlined the game's rules. The lender would place one white and one black pebble, and would request the girl to pick one.

If it was white, the loan would be wiped out, and the girl would not have to marry the lender. If it was black, then the loan would be wiped out; however, the girl need not marry. To win by hook or crook, the lender picks up 2 black pebbles and places them in the covered bowl. The girl, sensing the foul-play, and also to ensure she would not fall into a trap, picked up the first pebble, and as soon as she realized that it was black, she dropped the pebble "accidentally" into the old bowl, which had all the other pebbles.

She apologizes instantly for the accident, and requests permission to see the remaining pebble to resolve the color confusion of the first one as white. So, that way, she got her father's debt wiped out, as well as stopping the lender from marrying her.

This story teaches us that threat, and the hunger to win, always helps in coming up with an innovative idea to overcome tough situations.

Design Thinking – This is one of the modern age *creative strategies* that draws upon logic, imagination, intuition, and systemic reasoning, to explore possibilities of what could be, and to create desired outcomes that benefit the end customer. It's a *discovery based learning* to achieve the desired solution through *divergent thinking*. This is one of the tools that should be brought into practice more often to change the mind-set of the team.

Collective Genius – When the Google infrastructure group was planning to launch a platform for Gmail and YouTube, there were teams of capable designers and data scientists who tirelessly built two solutions: one on Big Table, and the other one from scratch on the current technology. Both the teams were asked to be honest, as well as to try to come up with a *spontaneous bag of options* to bump up the idea on which they were working. After a healthy brainstorming of over two years, it was quite clear that the option of building from scratch on the current system was not future-proof, though it may have solved the immediate problem at hand. The whole project was stopped after 2 years, and the Big Table team continued to build the platform. However, some of the key folks from the other teams were moved to the next project on hand for Google to lead the innovation. To summarize, apart from deciding the right platform, Google achieved in *creating a few more innovators* who could lead the next set of innovations, again and again. While delivering solutions for clients, exercising collective genius with more options, conventional and unconventional, helps both the client and the supplier organization.

Is frugal innovation an answer to this? Is creative diversity exploited fully?

Let's look at 2 such stories on frugal innovation and creative thinking, which are grounds up and have positively influenced many lives.

David – The Giant Killer

3,000 years ago, when the Kingdom of Israel was in its infancy, along its eastern border, there's a mountain range. And in the mountain range are all of the ancient cities of that region: Jerusalem, Bethlehem, and Hebron. And then there's a coastal plain along the Mediterranean, where Tel Aviv is now. And connecting the mountain range with the coastal plain is an area called the Shephelah, which is a series of valleys and ridges that run east to west. It's gorgeous, with forests of oak, wheat fields, and vineyards.

Then there were a group of seafaring people, originally from Crete, called Philistines. Both the groups confronted each other in a decisive battle in the Valley of Elah.

The Israelites dig in along the northern ridge, and the Philistines dig in along the southern ridge, and the two armies just sit there for weeks and stare at each other, because they're deadlocked. Neither can initiate first strike, as whoever does it has to run down the mountain to the valley, and would be completely exposed to the enemy.

Finally, to break the deadlock, the Philistines proposed for single combat, as it was known in ancient warfare, where one person from each of the groups would fight, and the winner would take all, and the loser would die.

The Philistines sent their mightiest warrior, a giant, who was armored from top to toe, with a long sword and a sharp spear. The Israeli army was terrified and was contemplating who to send on a suicidal mission. Then, a young shepherd boy came, and he asked for permission from King Saul of Israel to fight the Philistine Denisovan. He had assured the king that he had protected his flocks from lions and wolves in the past; hence, he should not be taken lightly. The shepherd boy refused to wear any armor; however, he took his stick, picked up five stones, and ran to the epic battle.

Looking at the response from the Israelis, the Philistines were puzzled; the warrior got wild, as he felt insulted. As the shepherd boy came closer, he took one of his stones out of his pocket, put it in his sling, rolled it around, and let it fly. The stones from the valley were not normal stones. They were Barium Sulphate stones. From medieval tapestries, as we know, the slingers were capable of hitting birds in flight. They were incredibly accurate. So, as the stone hit the giant, it was equivalent to a bullet from a 0.45-caliber handgun. The stone hit him right between the eyes, where there is no armor—in his most vulnerable spot—and he fell down, either dead or unconscious. The shepherd boy ran up, took his sword, and cut off his head.

The Philistines fled from the war.

This is an incredible story, which explains that the silver bullet can come from anybody within the team. Respect and nurture such brave souls to be creative all the time. Secondly, if you play to your strength, and utilize your local knowledge and keep it simple, then like the startups, Fintechs, anybody can be a giant killer.

Gandhian Engineering – Value for Many – More from Less for Many

It's about a master craftsman from Jaipur, City of Palaces, in India. Mr. Ram Chander Sharma had a humble beginning in a family of sculptures, where he grew up watching stones and metals being chiseled to life-size replicas of warriors. He was working as an arts and crafts teacher in Jaipur, when he was called by an orthopedic surgeon, Dr. P.K.Sethi, to teach disabled children, in Jaipur Sawai Man Singh Hospital (SMS).

During his tenure, he was intrigued by a proposition from Dr. P.K Sethi to manufacture light, durable, low-cost prosthetic legs. Another challenge at hand was to create something that is user friendly in Indian conditions, and useful in India ways of life, like walking barefoot, sitting on the ground with folded legs, or climbing trees with bare feet.

What he created from aluminum and vulcanized rubber (discarded car tires), has gone into history books as one of the finest frugal inventions, meant for 4 billion people in the world who earn less than 2 USD per day. The custom made foot is made on site and completely with limited resources; it costs 28 USD, compared to the bionic foot in the west, which costs 20,000 USD.

So an innovation need not always be high cost, or frugal innovation does not necessarily mean low cost. Rather, an innovation, if it can do more from less, for many customers, is a worthy idea to pursue—like how Amazon started.

We have covered customer centricity extensively, followed by how vital communication is in partnership. In this chapter, we discussed the creativity aspect in a relationship. But the heart and soul of any engagement between two parties is the extent of commitment from both. Don't miss the nice story from the Middle East, in the next chapter.

My 3 Cents

1. Create a team of creative minds, and challenge and incentivize to come up with new ideas that can be shared with the buyer organization.
2. If possible, try to infuse diversity in every performing team to take it to an even better team.
3. Be transparent with the buyer, on the resourcing strategy on improving diversity. These discussions should be with the senior management from all three types of buyers: business, IT, and sourcing.

CHAPTER 8

Commitment

"The attitude of giving a full commitment to the partnership will usually result in getting the same commitment in return."
– Denise Morrison

Commitment in a relationship is paramount, and demonstrates the longevity and quality of the partnership. Be it a personal, social, political, or business relationship, the start is often transactional or tactical. The expectation is fulfilling the immediate ask. However, if the motive is to translate the success of the transaction into a repeatable engagement, the bare minimum expectation from the buyer is to see the commitment of the supplier.

As the relationship moves more from the *forming* stage to the *performing* stage, or the growth stage, expectation from the supplier increases multifold. As a supplier, that's the right to latch on to these opportunities and move up the value chain.

Commitment is an assurance provided by both parties to each other through a series of tangible actions, which can help each other grow and also safe-guard in adverse market conditions. It embodies trust, transparency, and sharing. Within two human beings, it takes time to reach the stage where both are committed to the relationship. Where the relationship is between two organizations, or representatives of organizations bound through a contractual agreement, not only does it takes time, it needs growth potential for both the parties.

Success is eminent if in a team; every player, or most of the players, play up to the best of their abilities, and up to 100% of their potential. You might not win all the games; however, you will start winning soon.

> *"You need a commitment that is long term and a commitment to leadership, because that's the only way you build excellence."*
> **– Azim Premji**

How can a supplier demonstrate its commitment to the partnership?

The supplier has to put itself in the shoes of the client and take some of the responsibility. In business terms, it's often referred to as *putting skin in the game.* Let's look at the objective of the buyer organization again, and identify where the supplier can make a difference. A helping hand, or an intent to help in any of these areas, will go a long way in terms of proving the supplier's commitment; and hence, the buyer can see a new partnership evolving.

1. Intellectual Investment
   a. Thought leadership
   b. Leveraging alliance and partners
2. Capability Investment
   a. Capability development
   b. Capability up-skilling – Provisioning continuous learning
   c. Infrastructure provisioning
3. Innovations
   a. Process
   b. Technology
      i. Fin Techs and Reg Techs
      ii. New solution/components

In a true partnership, both the partners share the accountability to keep each other motivated to run the extra mile, which is beneficial for the business. As Simon Sinek says, *"Accountability partners keep you committed all the time."* Even if one loses motivation momentarily, the true accountability partner pulls one up.

"Individual commitment to a group effort – that is what makes a team work, a company work, a society work, a civilization work."
– Vince Lombardi

On that note, let's discuss the first aspect of any relationship or partnership. Investing in it is the most important step in the journey. Rather than being poetic or romantic here, let's delve into the different types of investments that are relevant for IT service suppliers.

The first type of investment is *Thought Leadership*. The term was first coined by John Kurtsman. It's the supplier organization's collective intellectual capital, which is relevant for the buyer to achieve their short-term or long-term goals. This type of artefact is mostly created by pre-sales or research teams in the supplier organization. It's a point of view (POV) or a *white paper* based on expertise within the firm, and experience across the industry. There is no dearth of relevant subjects at any given point of time. This type of investment does not need any additional effort on the supplier's part, as this intellectual property, created once, can be shared with an entire customer base, which includes the buyer and its competitors as well.

The second type of intellectual investment is the supplier's capability to bring in its alliances and partners to the foray, to help the buyer realize one or more goals. There is a possibility that the buyer is already engaged with some of the partners directly for some transactional sales. As a service provider, the supplier can still join the dots by leveraging their own alliances to suggest or come up with a unique proposition, or some proposition that has worked with competitors of the buyer. This way, the supplier is investing time and effort on behalf of the buyer to address some of the challenges, tactical or strategic.

This type of investment needs creativity and solution centric thinking at the supplier's end. An outsider's perspective sometimes opens up a new set of possibilities in resolving a client's problem. Often, in the day-to-day regular engagement, both customer's and supplier's thinking gets aligned. All the constraints get aligned, and the problem solving capability in the partnership gets weakened.

A story from the Middle East

Long ago, there was a wise old man, with his three sons, in a deserted village in the middle of a desert. He had 17 camels, which he used to rent out as a means of shipping in the desert. He had a hard life; however, he was able to make enough money for his family to have a good life. His sons had taken the life for granted. The old man was not happy with the attitude of his sons. He created a will for all his assets; however, he made sure that his sons should inherit only if they have enough commitment to solve the problem.

One day, he passed away, and after the funeral formalities were over, the three sons looked at the will. Their father had divided all the property into three parts. However, he had divided the 17 camels in a different way. They were not shared equally.

The old man had stated that the eldest son would own half of the camels, the middle one would get one-third, and the youngest one would get his share of camels as one-ninth!

The sons were surprised and were equally puzzled, as 17 is a prime number and cannot be divided. Several days passed on!! The sons were trying hard to solve the riddle, and were not able to really make use of the camels for their benefit.

After a few months, a new nomad family visited the village and were hosted by the family of three sons. The family was headed by an octogenarian granny. She was quite happy with the hospitality, and offered them a favor in return. The sons were not sure what to ask for. The elder son shared the ongoing problem of their father's will.

To their utmost surprise, she promised them a solution. She offered them one of their camels as a gift, if that could solve their problem.

To everybody's surprise, the offer provided a fresh perspective, and now they had 18 camels. With 18 camels, the elder one's share came out as 9, the middle one's share as 6, and the remaining 2 camels were for the third son. Altogether, it summed up to 17!!

The solitary camel, which was gifted by the old lady, was returned back to her.

This shows how the supplier could be the *old lady* in the partnership, and invest one and the most key camels to provide a completely new perspective to solve the customer's problem.

> *"Unless commitment is made, there are only promises and hopes...but no plans."*
> – Peter Drucker

Chetna is a Fellow of Yale, and a Schwab and an Ashoka, and the winner of the Forbes India Leadership Award 2017, as the *Entrepreneur with Social Impact*. Being part of an affluent family in India, influenced by Jai Parakash Narain, she got married, in a village, to an incredibly handsome man.

She was staying with my three children in the village. One day, a village woman called Kantabai came to her and asked if she could open a savings account. Chetna, bit surprised with the request, asked Kantabai, "You are doing the business of a blacksmith. Do you have enough money to save? You are staying on the street. Can you save?" Kantabai was insistent. She explained why she wants to save so that she can buy a plastic sheet before the monsoons arrives. That will save her family from monsoon. Chetna went with Kantabai to the bank. Kantabai wanted to save 10 rupees a day—less than 15 cents. The bank manager refused to open the account for Kantabai. He said Kantabai's amount was too small, and that it wasn't worth his time. Kantabai was not asking for any loan from the bank. She was not asking for any subsidy or grant from the government. What she was asking was to have a safe place to save her hard-earned money. And that was her right. And Chetna thought, if banks are not opening accounts for people like Kantabai, why not start a bank that will give opportunities for women, like Kantabai, to save? Chetna applied for the banking license, to the Reserve Bank of India.

The rest is history!!

The Reserve Bank said that they could not issue a license to a bank that is promoting members who are non-literate. Chetna was upset and she told Kantabai and the other women that we couldn't get the license because we are non-literate. The women said, "Stop crying. We will learn to read and write, and we will apply again; so what?"

The women started literacy classes. Every day, the women would come. They were so determined that after working the whole day, they would come to the class and learn to read and write. After five months, we applied again, but this time I didn't go alone. Fifteen women accompanied me to the Reserve Bank of India. The women told the officer of the Reserve Bank, "You rejected the license because we cannot read and write. You rejected the license because we are non-literate." And they said, "There were no schools when we were growing up, so we are not responsible for our non-education." And they said, "We cannot read and write, but we can count." And they challenged the officer. "Then tell us to calculate the interest of any principal amount."

"If we are unable to do it, don't give us a license. Tell your officers to do it without a calculator and see who can calculate faster."

Needless to say, Ms. Chetna Sinha got the banking license and started *Mann Deshi Mahila Sahkari* Bank. It's a microfinance bank.

Today, more than 100,000 women are banking with this firm, and have more than 20 million dollars of capital. This is all women's savings and women's capital, and no outside investors asking for a business plan.

> *"My courage is my capital. And I say here, their courage is my capital."*
> – Mrs. Chetna Gala Sinha

Innovation is the buzz-word, and it challenges the intelligentsia on both the buyer and the supplier side. Time and again, in the book, I have highlighted the importance of those baby steps of incremental creativity to bring in innovation at every interaction or *moment of truth*. Some of the initiatives, you would all be leading already, like Innovation Fund, or Innovation Labs. As long as these initiatives are aligned to deliver certain goals as per the buyer strategy, through joining different solutions provided by partners or fintechs, or internally developed, the utility as a supplier will be appreciated by the buyer.

The next category of investment is tangible, and it involves capital and significant effort at the supplier's end to ensure the initiation and effectiveness. This investment is around pro-active capability, building to achieve short to mid-term goals, as well as continuous capability upliftment to ensure that the skill set of the workforce supporting the buyer's organization is relevant for long-term goals. That takes us to the next chapter.

One of the few tangibles in the IT service industry is the human resources. The employees, or human resource capital, is the most important stakeholder in the whole scheme of things here. We will discuss the next "C"—capability—in detail. This aspect is responsible for the outcome of the service being provided to the customer by the supplier.

My 3 Cents

1. Create a charter on strategic investment initiatives and share with the buyer.
2. Share your big bets and request how buyer can help.
3. Create periodic plan for capability development and initiate dialogue with buyer for on volume commitment or at least acknowledgement whether it's in line with buyer's goals.

CHAPTER

Capability

"Man often becomes what he believes himself to be. If I keep on saying to myself that I cannot do a certain thing, it is possible that I may end by really becoming incapable of doing it. On the contrary, if I have the belief that I can do it, I shall surely acquire the capacity to do it, even if I may not have it at the beginning."
– Mahatma Gandhi

The Great Mahatma Gandhi, known for his mantra of *non-violence*, used it as the weapon for his successful bid for India's independence from the mighty British rule. He built the capability with whatever competency and capacity were available at his disposal, to achieve something not short of a miracle, back in 1947.

Competency – *Sufficiency of knowledge, skill, and strength to satisfy the wants of life.* This is what an individual brings to the table, at the outset, to play a particular role in a subject. The extent can be measured based on an individual's strength. While organizations recruit, it's a two-stage assessment. Firstly, "Is he/she, or not?" If the answer is *yes*, then comes, "How good is he/she in the current context, and for the future?"

Capacity – *The ability to receive, contain, and then perform, yield, or withstand.* It's the ability to learn or gain knowledge, to improve competency and use it to the best of its ability, and produce maximum output without affecting the quality. This is applicable for both machines and human resources. So the overall output can be controlled through adding or reducing capacity. In the case of human resources, unlike machines, consultants with similar competency from a supplier organization can be brought on board as additional capacity for execution.

Capability – *The collection of different competencies to fire at full capacity at the need of hour to complete a task efficiently and successfully.* Let's look at the following 2 examples, one in individual sport, and the second one in team sport.

Is Nadal capable of beating Federer, and vice-versa? Yes, both have different skills; however, both are competent and can play at the highest level of intensity for a long period of time. But how about the rest of the tennis players seeded outside the top 5? Unless Federer or Nadal is having a bad day, where they are either struggling in capacity or stringing all the skills or competency, the opponents are not capable of beating them.

Greece started the 2004 Euro Football Championship as one of the lowest ranked minnows. To everyone's surprise, Greece beat Portugal in the final, to win the cup and create history. Greece was ranked 14 of 34, within 4 weeks, in July 2004. Greece earned the respect of the football world, as on their day, the team was capable of beating the world's best. However, since 2012, Greece could never replicate the same kind of success. On the contrary, France, Italy, Germany, Brazil, and Argentina are some of the football teams that, in spite of their eventual scorecard for a championship, have the capability to win.

It's primarily because *capability* is the collective abilities of a group within a defined process to maximize the throughput of each competency. In spite of a change of players over a period of time, the process ensures the quality does not drop, and the outcome is unbiased. Improving capability is feasible in a short time, as opposed to creating clones of competent resources for increasing capacity.

Both competency and capacity building can be achieved through training and recruitment. As these two categories are more individually centric, and around 20% attrition is inevitable, the process is linear, and it is easy to maintain the quality and quantity of the workforce. On the other hand, capability development is more rounded, and often cumbersome but sustainable.

The good news is that capability can be built, uplifted, and changed with the vision or strategy.

Thinking from the buyer's perspective, competency and capacity development are part of hygiene. To be a strategic partner, the supplier has to devise a strategy for capability development, covering the following principles. The details can be different from one buyer to another; however, the principles remain the same.

1. What to build by when – This covers the set of actions, like recruitment and training, which the supplier is planning to take in the next 6 to 18 months to build the capability around the key programs and initiatives of the buyer.
2. How to build –
    a. Competency and Capacity – Internal – Supplier internal capability building will be on schedule for recruitments, as well as for the tools and frameworks necessary for training and re-skilling new and existing resources to deliver the capability committed.
    b. Competency and Capacity – External – How to Leverage Alliance and Partners – Supplier has committed to certain plans and investments. To ensure the effectiveness, as well as spend, the supplier should leverage the alliances and partners for building the capability. If that requires the buyer's intervention, it should be shared with the buyer for closure. Secondly, any new alliances or partnerships with university or government organizations should also be explored, if that helps.
3. How to Build – Capability – Supplier should create avenues for the competency and capacity resources to understand.
    a. Basic details of the buyer's program or initiative
    b. Relevance of the program to end customer
    c. Importance of their contribution to the success of the initiative
    d. What approach is taken (explained in next point)
4. Which Approach – New or Renew – To maximize the throughput and also bring in innovation to the mix.
    a. Re-invent the processes, with a touch of creativity to improve efficiency.

    b. Bring in new offerings to reduce dependency on human resources.
5. Why to Build – To maintain coherence – The supplier should share the view point with the team so that the team understands the necessity to keep the balance between diversity and focus, and between innovation and stability, as committed to the buyer.

As a supplier, the expectation is to fulfil all the reactive demands, both for human resources and infrastructure. Those demands can also be addressed by the people providers; however, there is no certainty on the price or location. This is where the buyers can make a difference, if the sourcing strategy is being shared with the supplier up-front, in the beginning of the year, or with sufficient lead time.

If you remember the Bullwhip effect discussed in previous chapters, if the communication between the buyer and the supplier is timely, then the supplier can ensure that the capability is available when it's needed. If the demand is higher or lower than forecast, then the supplier can only be bothered to manage the minor fluctuations rather than provisioning the backlog or surplus capability.

> *"Employees don't need to be best friends, but there does need to be a level of mutual respect and understanding."*
> **– Kathryn Minshew**

A bit of consideration and discipline from the buyer can work for both; even the supplier will be happy to go the extra mile for the next category, which is *capability upliftment*.

Automation – Is it a threat or opportunity?

Employees are assets—not costs and not machines. They are the most important stakeholders. Unlike products or machines, the quality, productivity, and throughput of human resources cannot be predicted easily. However, the good news is, with every new invention or automation, human beings are resilient enough to learn and adapt to new circumstances over a period of time. Learnability is the greatest asset we humans have. Nobody likes change, so whenever we encounter a new thing to learn or to master, we struggle in the beginning. It's inevitable that technology will replace a lot of jobs, even regular jobs. Not only in the production industry, but even office workers are in jeopardy, and might be replaced by robots, artificial intelligence, big data, or automation.

Let's take the automotive industry. More than 40% of industrial robots are already working, and automation has already taken place. Around 10% of car parts were electronic, back in 1980. Today, this number is more than 30%, and by 2030, it's predicted to reach 50%. With the new and innovative electronic parts and applications, we require new skills, like the cognitive systems engineer who understands and optimizes the interaction between driver and electronic system.

Have you heard about 3D printers and the Samsung Techwin Pick and Place machine? This is a classic, new age, *factory in a box*. This machine can put out 23,000 components per hour, onto an electronics board. This is the future of manufacturing, where the cost of innovation, prototyping, distribution, manufacturing, and hardware is getting so low that innovation is being pushed to the edges, and students and startups are being able to build it. This is a recent thing, but this will happen, and this will change just like it did with software. Now this machine will influence some redundancy of old skills; at the same time, it opens a whole lot more skill sets, to build, run, maintain, and enhance, and to take it to the next level.

If an apple-to-apple comparison is done, 70% of the skills present today were not there in the pre-Internet era. The same pattern can be predicted, due to the adoption of automation in the last 30 years, and some of the jobs have been obsolete as machines can do them; however, there is a flux of new skill sets that have evolved, which were non-existent then. Both factors are going to impact the landscape of re-skilled, highly-skilled, or newly-skilled resources available in the next 10 to 12 years. There is another factor that is going to be influential, which is the population growth in the current era.

Employees are the most important stakeholders. Unlike products or machines, these resources cannot be cloned or controlled in a tangible way. So the supplier has to be careful in the following four key areas:

1. Be on top of the demand forecast and supply, so that the Bullwhip effect is minimized.
2. Invest on attracting skilled resources.
3. Predict the future trends, and invest in proactive re-skilling and up-skilling of a few key areas.
4. Be flexible in retaining skilled resources.

With that, let's see what's being covered in the next chapter.
Not one of us is smarter than all of us!
That takes us to the next "C": collaboration. Time is short, and the changes in the technology world are so fast that one cannot learn everything and create competence in every field. Hence, we should collaborate more and more to be smarter, faster, and more innovative and efficient.

My 3 Cents

1. Every year, or on a periodic basis, please discuss the demand forecast with the buyer, and share the plan on preparedness of capability availability.
2. Track the wastage, prepare a periodic report, and share with the sourcing buyer.
3. Identify 3 key skill sets that are relevant for the buyer's objective, and invest in capability building.

CHAPTER 10

Collaboration

"Alliances and partnerships produce stability when they reflect realities and interests."
– Stephen Kinzer

Collaboration is an omnipotent tool; if used in the right spirit, it can yield wonders. The classic contemporary examples are Uber, Android, YouTube, Airbnb, etc. Collaboration is a proactive or voluntary intent to collect or leverage and share information that can be beneficial for all the parties involved.

Collaboration is like a placebo that enhances the overall throughput and individual productivity of a group of competent individuals. It's a multiplier, as long as there is no conflict of interest. Unfortunately, in every collaboration, there will be conflicts; and conflicts should be managed amicably before it destroys the whole fabric of collaborative success.

Red Zone and Green Zone Chickens – Teaching us Collaboration and Conflict

In Purdue University, a group of professors were experimenting on a group of chickens to assess the impact of collaboration and conflict. There were individual performers in the group, who used to lay eggs the most. The rest of them were average performers, and the productivity and quality of the eggs laid by those chickens were low in numbers, and average in quality. However, to their amazement, they noticed that the high performing chickens were quite aggressive.

When they stayed with an average performer, they would bully those average performers by picking on them all the time; hence, their productivity would be lower than average.

So, as an experimentation, all the average ones were kept in one cage and had been named the Green Zone Chickens, and the high performers were kept in a separate cage and were named the Red Zone Chickens. After four weeks, the results were astounding. For the Green Zone Chickens, the productivity had increased dramatically. At the same time, the chickens in the Red Zone were badly injured, and the overall behavior was detrimental for the group. As a result, the egg production had gone down significantly.

The learnings from this exercise are as follows:

1. In a conducive environment, with the highest level of collaboration, even without any individual star performer, throughput can be of the highest standard.
2. In a hostile environment, due to internal conflicts, a group of bright individuals can self-destroy.

Now let's talk about conflict. Is it good or bad, or inevitable? Conflicts arise when passions clash, and independent viewpoints contradict each other. So, it's not bad all the time, as it breeds mostly due to creative differences. If conflicts can be managed appropriately, the output can be revolutionary or transformational. Safety or mediocrity maintains harmony; however, it does not allow novelty.

On that note, let's discuss the different types of collaborations, where a supplier can leverage and avoid mediocrity.

There are four types of collaborations.

Internal – Resource based

As one of the old sayings go, "Wealth shared becomes half; knowledge shared becomes double." Collaboration encourages sharing knowledge based on individual competency and capacity. One additional catalyst that decides the effectiveness of collaboration is the *Extra Miler*, as described by Ning Li, at Iowa University. It's the team member or employee who has the mind set and desire to go the extra mile to gather and share information for the betterment of the team's outcome. So let's call them the *collaboration catalyst*. There are 3 types of collaborative resources Informational – These resources are vigilant ones. Apart from a regular 8 to 5 job, they always collect additional information, and share. They might not even realize the value of their input. They are the gatherer of missing or misplaced parts of a puzzle or a bigger picture.

Social

These resources leverage their network and access to certain information due to their position in the team. They can also bring in certain information from outside the team remit.

Personal

These are the competent resources, which can also be informational or social, and the quality of information collected by these resources are often contextual, as they are capable of joining the dots before sharing. However, they are always short of bandwidth due to their usefulness in the success of the team. They are always in demand.

The study says that around 30% of the benefit of collaboration is made possible by these minority (4%) groups of resources. The more extra milers we have in a particular relationship, the better collaborative index the team will have. So, to be different and efficient, the supplier should invest in these kinds of resources as part of the mix. There are a lot of artefacts available, which can help organizations to improve the heterogeneity in the mix, as well as the collaborative output internally.

Internal – Team based

The collaboration between different teams within an organization is often either mandated for an organization goal, or to come up with a creative solution as part of their response to a bid. The third category is where teams come together to create a generic solution, which can be offered to multiple clients as a service or utility.

Is there an extra mile we can stretch further? Yes, there is.

I have explained this as part of the capability building as well. As we have seen in Chapter 2, there are numerous touch points between the supplier and the buyer personnel. Can we leverage the internal team based collaboration, and speak the same language? Imagine the richness of those conversations. Of course, the quality, extent, and depth will vary depending on whether you are performing the role of informational, social, or personal. I have shared a few simple steps as part of the takeaways for your reference.

External – Partners and Alliances

This is one of the quotients where most of the IT service suppliers are doing well. Big/small; new/old; fintech/regtech—the only thing here that can be improved further is the awareness. Remember the warriors on the ground. The collaborative resources should be aware of the partnerships and solutions that can be leveraged in the remit of client operations. Secondly, the sales and delivery leadership can come up with unique propositions based on the strength of alliances, to bounce off as per joint planning, or based on need.

Sometimes this could cannibalize the solution if the partner-led solution is better suited. The transparency shown in adopting a partner-led solution always helps the supplier, being seen as an independent adviser to the customer.

Partnership or alliance forums are another set of avenues where, as a supplier, we can foresee the newer and path-breaking solutions available in the market. The exposure will provide multiple options to solve a business problem in the client's place. Hence, the quality of the supplier's response (proactive or reactive) will be rich and relevant.

External – With Buyer

That brings us to the fourth type of collaboration, and the most important one, as it involves the key stakeholder: the buyer.

Surprisingly, the other 3 types may, or can, generate new business; however, this helps the buyer in building individual brands, and helps the supplier to move closer to partnership. Also, do not try to run before you walk. Hence, the other 3 types of collaborations are inevitable before you succeed in this particular type. This category of collaboration demonstrates the extent to which a supplier is willing and able to bring in the best, by leveraging internal and external alliances, and directly involving the buyer's senior executives in the development of new offerings, products, or business solutions.

Success in this type of collaboration depends on the priority and intent of the buyers. All 3 categories of buyers can participate. This type of collaboration addresses the personal growth strategy for the buyer, as bare minimum.

Some of the notable examples are:

- Co-authoring an article with one of the buyer's key stakeholders
- Inviting the buyer representative to attend internal and external business forums
- Inviting the buyer representative to be the key note speaker in certain events
- Participating in one of the charity or social causes led or supported by the buyer

I would like to end this topic with a famous collaboration story in the tech world: CAPTCHA.

Louis Von Ahn, leading a massive program on digital revolution in publication in CMU, accidentally invented CAPTCHA, which we have been forced to verify on web to ensure the submission is done by a human being, and not by a robot or machine. This revolutionary idea paved a way for a startup named reCAPTCHA, which was bought by Google years later. The whole purpose of this idea was to improve security on the web; however, Louis felt that 500,000 hours of time, every day, is wasted by humanity as a whole. To ensure that the massive scale, online collaboration would be productive, he introduced this technology as a feeder to digitize millions of artefacts, so that it could be made available in electronic media, like e-books and Kindle, for words that OCR technology cannot decipher, or words from numerous languages where the OCR capability is not available.

Please pat yourself on the back, and take a bow. You should be proud of your achievement, and will continue to be so, for the next 15 years.

As of today, it's a huge success, where every human being—or at least 750 million of us—have contributed at least one word or one sentence to create Artificial Intelligence for the dumb machines, so that they can be effective enough to replace us!!

With this, we come to the last but not the least point for discussion. Excellence is not a destination; it's a journey. So that takes us to the next chapter, where we will delve into the importance of *continuing* the good practice.

My 3 Cents

1. Please identify those Extra Milers, and invest in retaining them in the team.
2. Brainstorm and create at least a couple of solutions per quarter, and take it to the buyer. From the initial traction and interest, you can gauge the eventuality of the propositions. Go with a *fail fast* approach: if it does not work, move ahead, and identify a few more.
3. Co-invest with a buyer in at least one initiative, which can improve the equity of the buyer.

CHAPTER 11

Continuity

*"Success is not final; failure is not fatal;
it is the courage to continue that counts."*
— Sir Winston Churchill

In the galaxy, universe, or solar system, or on Mother Earth, all are relevant, as long as there is a movement or continuity in the journey. As long as the water is flowing, it's fresh. The moment it stagnates; it loses the purpose. That's continuity of movement: momentum.

If the momentum is zero or negative, it results in an archetypical ending through a downward spiral. In the business world, it's catastrophic. It affects the morale and the output. Hence, it's quintessential to maintain a positive momentum for survival.

Momentum can be explained in science as Mass times Speed, with a direction, or in mathematical terms:

P (Momentum) = m (Mass) * v (Velocity = Speed vector)

So, to maintain a constant speed in the right direction, the supplier has to continuously put effort to negate the external forces, like attrition, and changes in geo-political, economy, or technology. That's continuity of momentum: perseverance.

As we are dealing with people here, momentum alone cannot help after a certain duration, due to the monotony. Whether it's a relationship between two individuals, or in group dynamics, maintaining the same level of commitment, productivity, and motivation is a challenge, as we are not machines. Hence, once on top, we will have to continue to do well to hold on to the lead or improve the lead. To improve the lead, the perseverance has to prevail with increasing momentum in the right direction. Then the momentum becomes a force to be reckoned with—a force that can grow with more mass and more speed. The same equation changes to

$$F (Force) = m (Mass) * a (Acceleration - Changes\ in\ Velocity)$$

That's continuity in improvement, commonly referred to as *continuous improvement*. It's the incremental improvement of a product or service over a period of time. Occasionally, a breakthrough improvement will come in with creativity and innovation; however, without the intention of incremental improvement, it's not possible. Rome was not built in a day. We could not have reached the moon or Mars without continuous improvement of the prototype by the Wright brothers, in the early 1900s. Kaizen is one of the well-known continuous improvement frame works, which helps immensely if followed regularly, and if action items are taken to closure. The IT service industry can take the learnings from the hospitality industry, or the aviation or health care industry, for that matter. Incremental improvements across sales, development, testing, or infrastructure will add up in increasing the momentum; hence, the supplier can create its niche, which can be a differentiator for a trusted partner.

> *"Excellence is a continuous process and not an accident."*
> **— A.P.J. Abdul Kalam**

Now, as we are on a positive spiral, and on a continuous improvement journey, what else do we need to raise in the eyes of the buyer, to be perceived as a potential partner? Any force or movement, good or bad, if not controlled, loses the relevancy. It's evident in the comparison between the nuclear bomb and the nuclear power station. That brings us to the next topic: continuous review.

- Is the force in the right direction?
- Is the force too much to sustain?
- Is the force relevant for a particular buyer?
- Is the force resulting in the desired throughput?

To answer all these questions, we will have to spend time on introspection and assessment of the current execution:

- We have created a joint plan as per the buyer's objective.
- We have put the communication framework in place.
- We have ensured that our offerings are creative.
- Our commitment to the relationship is in place.
- Our capability building initiative is doing well.
- We have covered the collaboration aspect to deliver value.

"It is paradoxical, yet true, to say that the more we know, the more ignorant we become in the absolute sense, for it is only through enlightenment that we become conscious of our limitations. Precisely one of the most gratifying results of intellectual evolution is the continuous opening up of new and greater prospects."
– Nikola Tesla

Should we wait for 12 months to review the overall health of a relationship? No. It should be instant or more frequent. Internal reviews are done diligently; however, the review by the buyer is more important. By the virtue of agility, DevOps, digital disruption, and the review and assessment by the buyer needs an instant makeover.

As I understand, the readers are esteemed leaders of the business world, and every firm has a defined assessment framework with parameters to do a health check. My two cents are around the periodicity, coverage, and relevance.

Periodicity – The frequency of feedback should be instant and fail fast. With every meeting, visit, presentation, proactive pitching, or proposal submission, we should seek feedback before we become the victim of the *out of sight, out of mind* syndrome. Quarterly and annual feedback frameworks are good, unless the buyer teams take a conscious effort to refer back to "What happened in the Quarters?" Secondly, when the duration is so long, there is a possibility of changes in teams on both sides. The third issue is quite prevalent: the recency effect. If in and around the annual appraisal period, in case of any adversity, the incident takes precedence over all the good stuff done in the last 12 months. More so for the supplier firm, the feedback exercise is not the true reflection of its performance.

Coverage – I have partially covered this aspect as part of the communication section. By coverage, I am alluding to the fact that most times, a supplier's success and good work does not reach out to the entire spectrum of buyer stakeholders. Hence, it's in the supplier's interest to share the success or good work across hierarchy and other relevant functions and units in the buyer's organization.

Relevance – A supplier aspiring to be a partner should prioritize this aspect. Year after year, the buyer's objectives change—same as the supplier's strategy to serve. So the template for assessment and parameters should undergo changes. It should be more contextual towards the buyer, so that while assessing, firstly, the buyer can assess based on their objective; and secondly, on the joint plan agreed before the start of the year. In that case, previous assessments, like instant, monthly, or quarterly feedback, can also be provided as reference. This way, the review will be effective, and assessment will be a fair reflection of the supplier's stake to partnership.

> *"Feedback is the breakfast of champions."*
> **– Ken Blanchard**

One of the greatest success stories on feedback

While the contemporary flood prevention systems in the world are designed to last for 200 years, the Dutch levee system is built to withstand the worst flood or storm, for 10,000 years to come.

Is it a path-breaking innovation, or carefully laid out, simple steps from past learnings, or both?

What's the problem statement here?

As the Greek geographer, Pytheas, writes about Heligoland, in 325 BC, or the Roman author, Pliny, mentions in his natural history, in 1st century BC: "More people died in struggle against water than the struggle against men." The Netherlands has been at the receiving end of the wrath of the North Sea—3 key rivers flowing across the country—for centuries. The northeast part of the land, which is called Nederland, is an alluvial plain below sea level, which has been built from over 2000 years of sediment deposits. The clay peat swamps were transformed to habitable land, with 350 miles of dykes as a sea defense system around the Northwest. The incremental innovation had provided short term relief till 1916.

A series of floods and storms, between 1916 and 1953, raised the need of a more future-proof solution. The Dutch engineers took a fresh approach to deal with the problem, and in the process, they could build a near perfect system, which every other country in the world is trying to emulate.

1. Big Bet – The objective was not to look at a short-term or a mid-term solution, but rather something that can be there for generations to come.
2. Nature or Consumer Behavior – There is predictability of the *unpredictable*, as long as things are not forced upon. Rather than building a defense system against nature, the dyke system and canal system across the Netherlands have an appreciation of how rivers, oceans, and storms behave in disasters, or otherwise. There is enough room for the water level to increase and decrease in adversity, without breaching.
3. Feedback and Continuous Monitoring – Millions of fibre optic sensors are laid out across the river beds, canal systems, levees, and dykes, in order to continuously monitor the parameters to predict the changing behavior, and take actions as necessary.
4. Notable Product Invention (talked about less) – A new fabric has been designed to hold the soil and stones from erosion. Though it was path-breaking, the importance is lost somewhere in the larger benefit of the whole project to humankind.

So, we need to keep the effort up. The intensity should not come down. We have discussed the 7 "C" principles in detail, and it's not a one-time activity.

A supplier has to be customer centric; however, the essence of customer centricity should change as per current relevance, the market, customer positioning in the market, and the need of the hour.

A supplier should keep the communication absolutely clear, effective, and transparent, and should be oriented towards the buyer's goals. Be a true advisor, like an accountant or a solicitor. Whether the buyer adopts, that's beyond one's control.

A supplier should challenge themselves to be both incrementally and continually creative. More so, it should be visible to the buyer, and also should help the buyer in improving the product offerings or services.

A supplier needs to demonstrate the commitment to the common cause. The investment should be worthy; this applies to both external customers and internal stakeholders.

A supplier should decide their positioning carefully. A service provider should build and provide capabilities, solutions, and frameworks with adequate cadence on delivered value articulation.

A supplier, who is smart and efficient enough to leverage a partner ecosystem and internal collaboration to create a packaged deal to solve one or more problem statements, will prevail long.

A proactive, trust-worthy, creative, collaborative, committed supplier has to continue the effort to challenge themselves, to sustain the good run through transparent communication, and ascertaining its unique selling point as a service provider. As long as the commitment and quality of execution capability are visible to the end customer or all categories of buyers…trust me, as a supplier, you have made all the right choices to be a worthy partner.

The Story of the Chinese Bamboo Tree

It's a famous story and is also part of most of the motivational speeches, blogs, and stories.

The Chinese Bamboo tree is a kind of tree that demands a lot of tender love and care during the initial days, or when the seeds are sown. The seeds are fertilized and watered for years. Yes, you read it right. It's nurtured continuously, every day for 1800 days, without any inkling of what's going on down under. But the farmers nurture with soil and water, and lots of patience. Once it comes out, it grows 90 feet tall in 5 weeks. That's an exceptional story of exponential growth in a short time. However, without the continuous effort being put in the first five years, the roots would not be strong, and they would not spread enough to sustain the rapid growth in later years.

Having said that, it's always the buyer's prerogative to choose the partner; nevertheless, the supplier's effort in bringing these changes will create a compelling case, which will force the buyer to think and act.

With this, let's move the focus towards the buyers. As the supplier, it's difficult to gauge and change the end customer affinity; however, we can still change the intermediate buyer's behavior. Let's look at the subtle adjustments that can positively influence the relationship with the suppliers, and behold!! It can help the buyers immensely to have partners delivering the service, rather than random suppliers. Let's move to the next chapter.

My 3 Cents

1. Keep on doing your best with your strength.
2. Share the view of value add and wastage as per the bullwhip method.
3. Put your stakes on partnership – Talk, cover, and communicate to buyers. Do not take "All good" as an answer.

CHAPTER

Sharing

*"If everyone is moving forward together,
then success takes care of itself."*
– Henry Ford

In any relationship, communication is vital for success. While two groups are in relationship, it is important that the goals are aligned, and that the goals are mutually beneficial. As we have seen in the success story of Toyota or Chrysler, the buyer has shared the long-term goal with the supplier, much in advance. That has helped both the buyer and supplier to bring their might behind the joint planning, which not only allowed them to fail fast; it helped them to have a realistic vision with a plan, which could help them realize the common goals.

Rather than being subjective, let's look at the aspects that the buyer can share with the supplier, in public domain or under a confidentiality contract.

- Organization Goals – The buyer's organization goals are often driven by business strategy. It's the final details of how business would like to perdition the organization in the marketplace from the consumer's perspective. These goals are often to be the best retail bank in a particular geographical area, or the most customer friendly retailer, or the best car manufacturer in luxury segment. Once these goals are shared with one or more suppliers in a formal forum, the suppliers can come up with strategies that can be adopted in a time frame, which has worked well for other customers or other buyers. If it is a unique goal, even then, the buyer can leverage door supplier's collective experience to come up with a new strategy. In IT services, the next set of details can be derived from the IT strategy and the sourcing strategy.

- Sourcing Strategy – The sourcing strategy helps in creating a pipeline of demands for a particular skill set, which would be required to deliver the IT programs necessary to achieve the business goal. To ensure the resources are available, the supplier and the buyer either have to build the capability, or recruit or free up the current available capability, working for other buyers or customers. It helps the supplier to create a plant demand and supply pipeline, which are analogues to the JIT (Just in Time) principle of inventory management. In the manufacturing industry, JIT is one of the most crucial principles, which helps both the buyer and the supplier to maintain a sufficient supply at any point in time; it's optimal to maximise production and to reduce inventory cost due to surplus supply. Unlike any other industry, in IT, the key supply is human resources, which cannot be recreated like a product or component; lead time is essential for the success of securing the necessary supply.

- Location Strategy – The services industry provides the flexibility to deploy themes in multiple locations. That brings us to the next dimension, which is called the location strategy. The location from where the service will be delivered is decided through multiple factors: where the end customer is, the location of the business stream, the location of the preferred IT team from the buyer's side, and lastly, the buyer's preference of the supplier's location. The other challenges for the supplier is their own growth strategy, in a particular location for a particular technology. At any given point in time, there will be more than 100 projects in the pipeline, which innately poses the next set of challenges: to either build a team or to move a team to a particular location based on the location strategy.

- Technology Strategy – The only thing that is constant in this world is this change, and it is so much more so in technology. It's exciting at the same time as it is unnerving to see the changes in the information technology space. With the uprising of mobile technology, user journey centric transformation, cloud strategy and various other digital disruptions, it's new norm for future. This throws up yet another challenge for skilled resources. But it's a boon. How does it impact the buyer and supplier relationship? The privy suppliers are in a business where they constantly de-skill and re-skill. They build the capability in the house; true training is recruiting from the market. Some technologies fail, some survive, and some thrive. It is not necessarily true that if one technology worked for a particular buyer, it would work for another buyer. But for the supplier, they have no choice other then maintaining the capability as long as it's serving the customers. Before we could build the capability in a particular technology, the buyer and supplier have to assess whether the new technology is fit for purpose. For that, either the buyer should leverage the supplier for multiple proof of concepts/points of view, or run some pilot projects to assess the fitment. This exercise is part of strategy building that should be finished before the joint plan is created.

- New Program Strategy – Last but not least, let's look at the strategy for execution. As part of the strategy, the item buyer and the sourcing buyer jointly decide the rules of engagement with suppliers, as well as the necessity of quantum of work to be packaged as a strategic program auto execute, in a tactical way through a business-as-usual work stream. This strategy throws up two sets of challenges.
    - Change in Process – Like new ways of working: agile, waterfall, DevOps, or CICD
    - Adoption of Tools and Framework – With the changing process, there comes a new set of guidelines for execution, along with a new framework, and lastly, a new set of tools.
  Adapting to change is one of the most difficult aspects for human beings. To achieve the outcome, it needs meticulous planning and relentless persuasion, and sometimes stricter guidelines for

adherence. Every organization has defined IT security policies, which safeguards the organization from malicious attacks on their network from the outer world. Every piece of hardware/software, or tools, goes through the baptism by fire, before implementation. That takes time and proving. This is one of the areas where the supplier's experience can come in handy, and it can help the buyer reinventing the wheel, as well as where the new tools and frameworks are concerned.

- Personal or Department KPIs – As we have seen in Chapter 2, the success of the buyer and seller relationship depends on the success of those thousands of touch-points during the journey. So, if I have to dissect the problem statement into piecemeal chunks, achieving success in every chunk is a feasible goal; and collectively, you can help by moving from a normal relationship to a partnership. That brings us to the last point, which the buyer can share with the supplier, and can seek help in achieving: the individual KPI/goal.

**"An expert knows all the answers, if you ask the right questions."
– Levi Strauss**

Let me introduce you to another case study on how objectives of both Buyer and Supplier can be common or complementary.

SCORE – Supplier Cost Reduction Effort

Chrysler Corporation kicked off a cost reduction initiative back in 1989. The primary objective for Chrysler Corporation was to achieve a significant reduction of cost in the manufacturing process.
Sounds Familiar!!
But the way Chrysler Corporation executed it demanded a lot of attention at the market. What did Chrysler do, and why was it a game-changer?

- Firstly, Chrysler invited 2,500 suppliers, briefed them about the objective, and told them how they could make a difference. The rest is history. Every supplier, or more so the primary suppliers, took this as a moral obligation to make Chrysler successful. The proposals submitted by the suppliers promised a savings of an astronomical figure in the range of 25 billion USD.

- Chrysler reduced the supplier base to 1140, from 2500.
- Selected suppliers were given the autonomy to come up with ideas to make it a win-win.
- Some of the suppliers, like Dodge, Dakota, or Textron Automotive, challenged their own teams. They came up with efficient product parts and cost effective ways to manufacture, so they could also gain some and share part of it back with Chrysler Corporation.

How did the supplier relationship change?

Parameter	Details	1989 – 1993	1994 onwards
Sourcing	Supplier selection	By competitive bid – Lowest price wins	By best design based capability for a set price
Accountability	For life cycle stages – design, prototype, and production	Split across different suppliers for different life cycle stages	Single supplier for end-to-end capability and delivery for a brand
Collaboration	Supplier investment in co-ordination and standardised practice	Minimal, as suppliers were working in silos and for lowest cost	Substantial investment in collaboration to improve the overall value for money
Contract	Duration	Short Term	Long Term
Assessment	Performance	Simple, cost based, and one way	Complex and value based, dual way; feedback from partner is expected
Recognition	Mutual Respect	Transaction Oriented	Relationship Oriented
Empathy	Supplier financials	No accountability towards supplier's profitability	Joint responsibility for lowering supplier's cost and improve productivity hence profitability
Commitment	Risk/Reward	Adversarial, zero-sum contacts	Collaborative, Trusting, recognition and rewards

How did Chrysler perform?

Chrysler's turnaround was noteworthy. Since the SCORE program was introduced, Chrysler's profit sky rocketed from almost breaking even, to more than 10%, by 1994. The rest is history. For more details, please visit my book website.

This is a fabulous example where the mission or objective of the buyer was translated into a group of tangible voluntary actions for the suppliers; hence, for the first time ever, the US automotive industry and US car manufacturers had 1140 partners who were keen to make the customer successful in the market.

> *"Collaboration is an important part of the process, and ego is never a part of it."*
> **– Mack Wilberg**

Another aspect of sharing in a partnership or team, is collaboration. The supplier collaboration, if managed positively, can lead to innovative ideas. At the same time, there is an inherent risk of supplier collaboration conflicting with the buyer's internal research and development. Also, supplier's need a bit of space and autonomy to bring in selfless collaboration. That's where the buyer can lead and show maturity by treating the supplier the same as internal resources. Collaboration breeds ideas, which may or may not result in a favorable outcome; hence, failure needs to be managed gracefully. Whoever is the leading partner, should demonstrate enough maturity to take the positives out of failure to practice a culture of comraderies rather than blame game.

That takes me to another great story of leadership. I am sure you will like it.

Abdul Kalam's story on sharing the Blame/Fame

On August 17th, 1979, in the Vikram Sarabhai Atomic Centre, the chairman of Indian Space Research Organisation (ISRO), Dr. Satish Dhawan, and the mission director, Dr. APJ Abdul Kalam, were preparing for a missiles launch program, where the vehicle is made indigenously. As the launch was initiated, at T-8 seconds, the computer took over the launch, and to everybody's surprise, the computer put it on hold at T-4 seconds, as it found there was a leakage on the second stage of the rocket system. Given the situation, where it's the first homemade rocket system, the ISRO team was under tremendous pressure to make it a success. Hundreds of engineers and scientists, under the leadership of Dr. APJ Kalam, had to make a decision within a minute. As the team informed that there was sufficient oxidiser in the second stage, a unanimous decision was taken to over-ride the computer and go for manual launch. The rocket was launched, and the first 100 seconds were spectacular, but then disaster struck. The second stage spun off mid-air and fell into the Bay of Bengal. After a few minutes, a press conference was called to brief the national media. To Dr. APJ Kalam's surprise, the chairman, Dr. Satish Dhawan, took responsibility for the failure, and addressed the press. At the same time, he was proud of Dr. APJ Kalam and the entire team for the maiden attempt.

Within a year, ISRO was ready with a second attempt. This time, the launch was a phenomenal success, without any untoward events. After 30 minutes, the entire team was preparing for the press conference. This time, the chairman, Dr. Satish Dhawan, insisted on Dr. APJ Kalam addressing the press and sharing the success with the entire world.

With this, let's move to the next chapter, where we will discuss the importance of engagement and, more so, the intent of investing in the business relationship.

My 3 Cents

1. At the beginning or the end of the year, organize a supplier forum, and present at least one business problem statement or an IT strategy. Request that the suppliers/partners share their view point on how to help in achieving the goal. After assessing the responses, pick one or two key partner's suggestions, fully or partially.

2. At the end of each year, or quarterly, request that the key supplier(s) share the wastage due to the bullwhip effect. This would help the supplier to be heard and be open to reduce the overall spend every year, with better quality of delivery.
3. Improve internal governance to improve the demand forecast, and share it at least before one quarter so that there is sufficient time for the supplier/partner to arrange/recruit/build better capability.

CHAPTER 13

Engage

"When you show deep empathy toward others, their defensive energy goes down, and positive energy replaces it. That's when you can get more creative in solving problems."
– Stephen Covey

When two firms get into a partnership or business engagement, like the IT buyer and supplier, one of the key aspects that matter the most for success is *how they engage*, rather than how frequently they engage. Let's look at a simple exercise below.

Still face experiment

At Harvard World Child Centre, the doctors did an interesting exercise on 7-month-old children. The mothers were instructed to turn away from the child and look back with a still face. The 7-month-old child has basic hard-wired knowledge and minimal instincts. When a child looks at the still face of the mother, at first the baby tries to get her attention by various sounds and gestures. After a few futile tries, the baby disengages.

So we, as human beings, need *continuous interactions*; and in a business context, it's far more frequent. Without engagement, the relationship loses purpose and, soon, disengagement creeps in.

Work force 2030

As explained earlier in the chapter highlighting the importance of capability building, 70% of the skills present today were not there in the pre-Internet era. Due to the adoption of automation and newer inventions over the last 30 years, some of the jobs have become obsolete, as machines can do them. At the same time, there is a flux of new skill sets evolved.

Thirdly, the current demography and rate of population growth will be another key factor in predicting the availability of re-skilled, highly-skilled, or newly skilled resources in the next 10 to 12 years.

A recent research says, by 2030, we will have a skill shortage on an epidemic scale; in spite of automation, it will reach 70%, compared to the current 20%.

As of 2020, the surplus skill availability in the top 15 economies:

1. Europe – Spain (-17%); the rest (7% or less)
2. Americas – USA (-10%); the rest (5% or less)
3. Asia – China (-7%); India (-6%); the rest (0% or less)

As of 2030, the view is quite sombre for the same set of countries, who control 70% of world GDP.

1. Europe – Overall, negative; for Russia and Germany (-20%)
2. America – US (-4%); the rest, all negative
3. Asia – India (-1%); the rest, all negative, including China

Overall, the picture is grim; hence, the current skilled resources and the new generation have to be handled with care. To maintain the growth, every firm would have to bid for the same set of surplus skilled workers in the world.

Both the buyer and supplier need a fool-proof people strategy, which has 3 key aspects:

1. A robust plan to forecast the demand for different jobs and different skills, well in advance, and shared with the supplier. 1.

Workforce planning will become more important than financial planning.
2. Create an environment conducive enough for the supplier personnel to feel part of the larger organization and be motivated enough to continue.
3. Provide enough freedom to the supplier to manage the skilled resource pool for you. In other words, stop micro-managing resources as the supplier is best placed to attract, retain, re-skill, and up-skill the talents for you.

Be it employee or supplier personnel, a bit of TLC always pays off.

> **"Empathy is the starting point for creating a community and taking action. It's the impetus for creating change."**
> **– Max Carver**

Empathy breeds trust and mutual respect. Unlike the tech buyers, like Apple, Microsoft, and Google, the buyers from other industries, like financial services, insurance, retail, energy, or utility, leverage the suppliers for advanced capability. The fact is, any IT professional, or IT aspirant, is not aiming to join a bank or retailer to enhance their IT career. The preference would be the product, consultancy, or services firms in IT. If we can keep the cost arbitrage aside, it's the experience and knowledge of the supplier personnel, working worldwide with all sorts of clients across industries, that makes them valuable for the buyer.

Let's look at some of the best practices buyers around the globe exercise, which demonstrates mutual well-being:

1. Buyer leads and facilitates Kaizen sessions to leverage supplier(s) point of view, to achieve the future goals (3 months, 12 months).
2. Buyer listens to the supplier(s) growth strategy for next 12–24 months, and the probable common themes for joint success.
3. Buyer is keen to discuss with the supplier and to empathize in certain conflicting goals.
4. Buyer is interested to know the well-being of the supplier's staff. Is the supplier's staff motived enough to work for the buyer?

5. Buyer spends sufficient time with the supplier teams to reassure their role in an extended team.
6. Buyer invests effort internally to share the strength of the supplier(s); hence, the perception is collective in nature.
7. Buyer encourages internal members to share feedback with the supplier, with maximum participation and objectivity.
8. Buyer is willing to take advise and feedback from the supplier.
9. Buyer acknowledges the internal operational challenges and takes corrective actions to de-risk the supplier's position for on-time delivery, governance, or timely payments.

Whether the base of the relationship is on cost reduction, value for money, innovation, or all of these together, involvement and acknowledging each other's position nurtures the relationship positively. Any negativity or perception should be dealt with as soon as possible before it brews into detachment.

That takes us to the next vital element: *empathy as respect*.

In the supplier partnership model, the buyer, being the senior partner, naturally assumes the control in the relationship. And suppliers will be fine with the arrangement, as long as there is mutual respect for the capability. But the buyer should remember that with control, comes the accountability. Control can be detrimental in the relationship if the buyer representative does not command respect, and if it is by the virtue of the buyer's position. The following are some of the steps that the buyer should take to extract the maximum from the supplier, and the supplier is happy to support the buyer in the pursuit of success.

- Understand the strength of the supplier personnel.
- If the supplier personnel are more knowledgeable, or is a subject matter expert, collaborate, seek advice, and involve them in decision making.
- If the supplier personnel or group is less aware, help them grow so that they can be contributing towards the buyer's success.
- The IT buyer or sourcing buyer should devise processes and train the supplier to follow. If the supplier has a better process, the buyer should evangelize roll-out of the practice within the buyer's

organization.
- Visit and spend time at the supplier's premises for the key stages of the program, to "live a day as the supplier," so that there will be appreciation of the supplier's effort, which at times could be black-box.
- Complement or develop compatible capabilities rather than competing.
- Invite the supplier or join the supplier in a capability building exercise.

A relationship is nurtured by the amount of time invested.

Time

In a geographically distributed team structure, does the buyer spend sufficient time with the extended team from the vendor, and vice-versa? From the vendor's perspective, is the team at the client site motivated enough? When there is an issue or a problem, are the teams on both sides enthusiastic to solve the issue first, or do they start a blame-game?

Time is the *most vital* investment in any relationship. In the pursuit of a goal, the amount of time and effort we put in determines the outcome. Irrespective of a romantic link-up or a professional relationship, both the parties have to invest in time for each other. Lack of investment in any joint venture slowly dies down over a period of time.

- What if the buyer has shared the vision, strategy, or goal with the supplier?
- What if the supplier has shared the plan to jointly achieve the goals?

Unless sufficient time is invested in each other, to assess the progress and take both proactive and reactive measures, success is hard to come by. Time spent together also signifies the importance and commitment from both the parties in the relationship.

Let's look at some of the examples or practices that can help strengthen the relationship between the buyer and the supplier:

1. The business buyer, IT buyer, and sourcing buyer should meet with the supplier representative periodically.
2. Even if the business buyer is not responsible for the supplier relationship, the business buyer should spend time with the key supplier, along with the IT buyer or sourcing buyer.
3. If the goals are shared with the supplier, both parties should spend time, fortnightly or monthly, to assess the progress of the KPIs, apart from the regular assessment parameters on quality and financials.
4. The buyer management should plan to spend time with the supplier teams, as that provides the feeling of comfort and inclusivity with the team.

Edinburgh Tram Project

The *inspiring capital*, Edinburgh, is an idyllic beauty by the shore of Firth of Forth, and is often regarded as one of the top 5 beautiful locations in the world. It is a world heritage site that attracts more than 20 million tourists a year, and it had everything, other than a tramline to provide another alternative in public transport.

Since the inception of this initiative in 2003, the project was in the limelight for political reasons. The premise was to connect the airport to the east most harbour—Ocean Terminal—crossing the heart of the city, to Princess street.

Items	Estimated	Actual
Route	Phase 1,2, & 3	Phase 1a (Curtailed to York Place)
Total Stretch	40 KM	14 KM
Start Date	2008	2008
End Date	2011	2014
Infrastructure and Other Costs	327 MGBP	714 MGBP
Vehicle Cost	48 MGBP	62 MGBP
Total Cost	375 MGBP	776 MGBP

Between the project inception, in 2017, and the real work, started in 2018, the budgeted numbers already went up by around 150 MGBP. Since then, the project performance went southward, till 2011, due to a lot of issues, like mismanagement, failure in communication between management and supplier teams, inadequate risk management, and unplanned delays due to environmental factors.

As per the enquiry report, and in the interest of our topic of discussion, let me summarize what things did not go well in this project:

- The sourcing process was rock-solid; however, execution was poor.
- Good practice and framework were in place; however, there was undue reliance of council on Transport Initiative Edinburgh(TIE), and lack of ability to deal with poor performance by TIE.
- There was lack of proper due diligence by council (buyer) and the deputed managing firm TIE, on the contracts with the suppliers, from a risk management perspective.
- There was a serious communication breakdown between TIE and suppliers in dispute.
- There was an error of judgment by TIE, by withholding payments for suppliers, and bringing a lack of trust to the equation.

The key takeaway from this case study:

- The business buyer (Council) and TIE (IT Buyer equivalent) should have done better due diligence.
- The business buyer should not blindly go with the assessment of the IT buyer, especially when the suppliers are more knowledgeable in the subject matter; rather, frameworks and good practices should be leveraged for better decision making.

How are we doing? We have covered the two key aspects: sharing and engaging. Unless we track the progress, any objective or goal cannot be achieved. With this, we come to the third aspect, from the buyer's perspective: assessment.

My 3 Cents

1. Buyers from each fraternity should plan for "a day in the supplier's life," per quarter. As a CIO, head of technology, sourcing manager, or business head, please plan and put this into practice. Even if not every quarter, at least start with a day every 6 months. Trust me, your gesture will work like magic with the supplier personnel's morale and productivity—like a magic potion.
2. Invite and engage suppliers worthy of being partners, for governance and strategic meetings. Whether to take suggestions from all or in entirety, is your call. The outcome would also be a testament of the supplier's capability to be a partner in the true sense, or not.
3. For any new initiatives, invite suggestions from the key suppliers. This would give you a head start, or pro-actively make you aware of any pitfalls or risks.

CHAPTER 14

Assessment

"I think it's very important to have a feedback loop, where you're constantly thinking about what you've done and how you could be doing it better. I think that's the single best piece of advice: constantly think about how you could be doing things better and questioning yourself."
– Elon Musk

Let's consider a real life example. I need my dream home: an independent house or bungalow, 2-storied, with garage and garden. I have 2 alternatives. Either I buy a house with all the features, or I build. The end customers or consumers are my wife and kids, who will spend most of their time in this house. In reality, I am the minority at home, surrounded by 3 ladies. So that explains my tricky customer base! Considering the business case here, I finally decided to build. I am the business buyer, with money. Again, I have two alternatives: either I buy a house with a plot, from a builder as part of their new development project launched, or I buy a piece of land and hire one of the best architects and builders to construct the dream house.

1. In the first scenario, I will have limited scope for changes, as the builder is going to use a cookie-cutter model; still, I can change a few things, like the layout of rooms, or add a new electric point, change the flooring from wood to carpet, etc. Depending on the appetite of the builder, further structural changes are possible; however, the outcome will be closer to our dream home, but

maybe not entirely.
2. The second scenario, I dictate the changes, and I can have whatever I want. With this comes the responsibility of ensuring my requirements are clearly articulated, and the builder acknowledges the understanding without any ambiguity.

One key aspect, which I should not be ignoring, is my periodic assessment of the project in terms of spend, progress, quality of work, and conflict or crisis management. Continuous feedback from me will definitely help the builder, even if their reputation is at stake. At the same time, I should take feedback from the builder on certain design considerations, as in hindsight, I might realize the non-usage of the facility due to pragmatic reasons.

That shows the importance of continuous assessment and two-way feedback in partnership.

In the IT services context, the sourcing buyer and the IT buyer are the closest to the performance of a strategic supplier. To gauge the value the buyers, get out of the relationship, there are numerous assessment mechanisms available with the buyer. One of the key tools used is the feedback mechanism. There are matrix and scorecards at the buyer's disposal to assess the supplier performance periodically. However, this does not provide a 360-degree view off the health of the partnership. That brings us to one of the key considerations: should we take feedback from the supplier?

> *"Life is the continuous adjustment of*
> *internal relations to external relations."*
> *– Herbert Spencer*

If the intent is continuous improvement of the relationship, two-way feedback always provides insight into some of the challenges that might be impacting the supplier's outcome in the long run. Nothing is more constant in the world than change. So the supplier feedback may highlight perceptions, like the supplier feeling disengaged, or the buyer strategy is not in line with the supplier's growth. More so, the gesture reiterates the importance of the supplier's role as even and equal in attaining the common goal.

It's natural that some of the feedback may not be constructive all the time. There is inherent risk of information sharing between two organizations, which might not be relevant or confidential. Depending on the maturity of the supplier personnel, the nature of feedback could be personal or incorrect due to a lack of awareness of the prime objectives after partnership. There will be apprehensions and nervousness around the finer details of exercising this practice. However, I strongly feel that the benefits will outweigh it, and will create a conducive environment for the partnership to flourish.

The most striking aspect of seeking feedback from the suppliers demonstrates the confidence of the buyer in the partnership. If the oddities in feedback can be used constructively, it would plug the gaps in communication with the suppliers, or the challenges the suppliers see on the ground while executing on a daily basis. There could be disgruntled suppliers or feedback full of inaccuracies; it's at the buyer's disposal to pick selectively and address for betterment of the relationship, if the supplier is relevant.

How many times have we heard that a team is out of governance, or there is delayed billing of resources? The Purchase Order is delayed or rejected. There is a huge back log of invoices to be paid. Does it all sound familiar? The overall governance is often a big issue for the suppliers; whereas, for the buyer, it could be irrelevant unless it affects their own key performance index(KPI).

So, the sourcing buyer or the IT buyer should be cognizant of the repercussions of these practices on the supplier's KPI. Remember the *bullwhip effect*, explained in Chapter 1. In a competitive scenario, every step in the procurement or supply chain, if the delay or wastage can be minimized through adequate forecasting, helps the suppliers in their profitability. If the supplier is profitable, in the spirit of partnership and keeping the relationship long-term, the supplier will be more proactive in passing on the benefit to client; hence, it would help the buyer in reducing the overall cost of ownership in IT.

As a senior partner or dominant party in the eco-system, the buyer has some accountability towards the relationship. In the same vein, another aspect demonstrates how much the buyer organization values the supplier. Every RFP or bidding soaks up enormous amounts of energy, effort, and bandwidth of participating suppliers. While seeking the bid, the dates are sacrosanct for the buyer. However, during the selection process, if a few things can be followed diligently, it exhibits the professionalism, as well as the seriousness, of the buyer towards the relationship.

- Stick to the dates published as part of the process.
- Or, proactively inform the suppliers.
- Each participant should at least be informed about the outcome.
- Share the feedback so that the supplier can learn from the exercise and serve better in future.
- Respect the supplier's time; hence, the outcome can ensure the bid is officially closed, and the team is made available for other pursuits.
- It affects the supplier's internal KPIs.

Let's look at a few simple and feasible steps. There are a few organizations who already follow these (partial) as part of an annual feedback exercise.

1. Informal feedback is taken from the partner supplier's management before the appraisal of the buyer, ensuring the feedback has no bearing on performance. The findings can be shared for self-awareness.
2. Formal feedback may not be at the personal level; it is more for the organization as a buyer. Again, the findings can be used selectively. If it's in line with the buyer's strategy for a certain supplier partnership, then that shows the success of the strategy, or else corrective action can be taken, if it's relevant for future strategy; or a simple dialogue with the supplier management will build the trust and inclusivity, beneficial for the relationship.
3. This can be applicable for all business, IT, and sourcing buyers.

That brings us to the next set of parameters on which a supplier should be assessed as a potential partner.

Performance:
- Financial
- Historical

Quality:
- Delivery
- Relationship
- Thought Leadership

Relevance:
- Automation
- Software as a Service
- Innovation

Compatibility and Trust Assessment (CaT)

Professors Mr. Karl Manrodt and Dr. Jerry Ledlow teamed together to come up with a compatibility and trust assessment framework, called CaT (Compatibility and Trust Assessment), which provides a 360-degree view of a business relationship and the teams involved in operating the relationship.

It's a two-way protocol and meant for both buyer and supplier. The framework allows both buyer and supplier to assess each organization's inherent tendencies and ways of working. This provides a global view of the relationship dynamics and alignments.

The framework looks at compatibility and cultural fit between buyer and supplier across these five dimensions: trust, innovation, communication, team orientation, and focus.

The outcome of the survey educates both the parties on:

1. Assessment of alignment when the relationship starts.
2. Perceived gaps and misalignment during the life-cycle.
3. Uncovering the probable future compatibility issues.

So, as long as there are avenues to maintain the assessments as two-way phenomenon, it acts as a proactive step to make the partnership a success.

Nothing would be more fulfilling then leveraging each other's strength to attain greater heights in the relationship.

There are more success stories together, more collaboration, more transparency, and more celebration. That brings us to the last chapter, where we will deliberate on the outcome of a successful partnership.

My 3 Cents

- Create an avenue where the key supplier(s)/partner can share independent and impartial viewpoints or feedback, anonymously. The information gathered should be used for constructive improvement in the buyer's behavior, and relationship with the partner.
- The sourcing buyer should assess the wastage due to the bullwhip effect, and share the information with the IT buyer and the business buyer. This will ensure that the internal process and bottlenecks can be improved.
- The business buyer should encourage transparent communication with the key IT suppliers. This will also provide an opportunity for the business buyers to assess the performance.

CHAPTER 15

Leading or Being Led

"The best partnerships aren't dependent on a mere common goal but on a shared path of equality, desire, and no small amount of passion."
– Sarah MacLean

Partnerships are successful when one leads all the way, or both lead on their individual strengths at times. So, be like Apple, Airbus, or Toyota, where these firms lead the suppliers, or be like Amazon or Google, where you leverage the strength of suppliers to flourish.

Play Hard!! Party Hard!!

Let's take a look at some of the notable, current age business partnerships.

Apple and Nike – Nike+

It's one of the best partnerships, where each provides a wonderful customer experience based on their individual strengths. Where the sportswear giant, Nike, provides the tracker, the app, from Apple, tracks the activity.

BMW and Louis Vuitton – Pure expression of the art of travel

This is where the high quality craftsmanship, from both the partners, is delivered in a co-branding campaign. The BMW i8 car, and the four-piece luggage that fits in it, are made of light weight carbon fiber, yet it does not compromise on the quality.

Success is a continuous journey; it's one way and non-linear, and it's circular. Hence, we cannot stop or relax on this journey. Then how can we sustain it? The only answer is to celebrate and move on. A successful partnership brings in new possibilities and newer challenges. It's contagious; so celebrate, praise, motivate, and look for a new frontier to challenge yourself and the team.

Like the economic climate, the whole technology world is going through a change for realists, and it's a transformation for the optimists. The consequences and repercussions of this change will affect the firms on both the buyer's as well as the supplier's sides. Hence, enjoy the present, share the success, nurture the investment in the relationship and, most importantly, celebrate the partnership.

> *"The whole life of man is but a point in time; let's enjoy it, therefore, while it lasts, and not spend it with no purpose."*
> **– Plutarch**

Any excellent recipe for food lies on a fundamental principle of maintaining a balance of ingredients and tastes, so that each ingredient contributes collaboratively. Creativity should surprise or please the consumer so that it ticks customer expectation, or forces the buyer to positively influence to accept the novelty in the dish. As long as the dish challenges the status quo and demonstrates the commitment of the chef to do something newer and better with both available resources, it adds to the success of the dish. Borrowing the quote from Knut Haanaes, and using the parlance here, in the relationship between a supplier and buyer, there has to be a balance between the dosage of exploration and exploitation.

Exploration is about finding a new frontier or promoting new ideas; whereas exploitation is making the good better and the better best. Exploration brings in a lot of idea sharing, innovation, and creativity; whereas exploitation ensures we do the same set of activities with a predictable quality outcome—but better, cheaper, and faster over a period of time. If the former is more for future and long term, the latter lives in present and short term.

Often, over a period of time in a long-term business relationship, too much focus on exploration leads to a perpetual search of the silver bullet, which is never ending; and the ideas die down before being implemented or tried out. At the same time, if there is too much focus on exploitation, we do the same thing, and there is an even greater danger of losing spark in the relationship.

> *"Give a man a fish and you feed him for a day.*
> *Teach a man to fish and you feed him for a lifetime."*
> **– Chinese Proverb**

Partners often play both the roles: sometimes feeding and sometimes teaching to feed.

With this, let's look at one of the most read and discussed case studies on supplier-led transformation, for one of the most successful auto makers in the world.

Toyota Story – JoJo – Slowly, Gradually, and Steadily

There are 7 key learnings from the Toyota case study below, which is absolutely relevant here, and it's up to the organizations in IT services to ignore their longevity.

- Long-term vision – from surviving to thriving
- Thinking beyond today – challenge the obvious
- Quality as collective responsibility – cannot be unilateral
- Play on your strength – buyer and supplier
- Use of efficient work practices across self and suppliers
- Optimum utilization of capacity and capability to make it a win-win for everybody
- Improve ROA (Return on Assets) and ROE (Return on Equities)

In July 2000, Toyota launched the CCC21 (Construction of Cost Competitiveness for the 21st Century) program. It's an overall 30% cost reduction initiative, which aims at deploying new concepts to reduce the price of 173 key components essential for next generation vehicles, without compromising on quality but still being the world's leading car manufacturer in the new millennia. It provided a challenging environment for all the key suppliers of Toyota, to think creatively and come up with solutions that are more efficient but built with lesser cost. The success of this program has redefined the relationship with the supplier. Contrary to popular belief, Toyota has found many partners who were happy to participate and achieve the goal for Toyota.

There were hundreds of new practices or best practices tried out, and it paid off. In the interest of our subject, let's look at the supplier relationship model: a vital differentiator.

- Keiretsu – The Japanese supplier partnership model – "Creating is a conglomerate of manufacturers, supply chain partners, distributors, and financiers who remain financially independent but work closely together to ensure each other's success." Along with Toyota, the group of suppliers were motivated to learn, improve, and prosper continuously.
- Toyota awarded long-term contracts, and encouraged key suppliers to play *big brother* for lower tier suppliers, for producing innovative integrated components or sub-systems as an accountable partner in quality, cost, and on-time delivery.
- Communicate and Collaborate – Toyota shared the vision and strategy with suppliers every year, and also heard their goals and strategies to support them in the journey. In this way, both Toyota and the supplier could jointly create a development plan to complement each other.
- According to Taiichi Ohno, the father of TPS (Toyota Production System), the parent company or the manufacturer would like to maximize profit; however, not at the expense of the supplier's well-being.
- Partnership of Equals – Toyota understood and helped key suppliers understand how process change can make them profitable; hence, they can reciprocate the favor through cutting their profit margin.
- Command in Control – Toyota believed in leading the partnership with a hands- on approach through continuous assessment, immediate resolution, and with intent of improving the outcome rather than being a hindrance for the supplier in achieving goals or resolving problems. If the manufacturer has the controlling role in the partnership, then they should be accountable for the

outcome; hence, a hands-on approach will help the supplier in adverse situations.

The success of the Japanese supplier partnership model was historic, and a game changer in the US automobile industry. A lot of research has been done on the case study of Toyota and Honda over the years. It also proved the partnership theory is not restricted to geography or demography, as both Toyota and Honda have proved the art of possibility in inculcating a new partnership culture with suppliers from US, Mexico, Europe, and Asia.

Let's end this example with an event in Toyota history, where Toyota rewarded the key partners around 100 million USD as a bonus for being so collaborative and savior when needed the most. It was on the morning of February 1, 1997. Fire broke out in one of the main factories of Toyota, which was used to produce brake fluid proportioning valves (P-valves) to help prevent skidding by controlling the pressure on rear brakes. These are used in the braking systems of all Toyota vehicles. This factory manufactured 99% of Toyota's P-valves, with Nisshin Kogyo Co. producing the remaining 1%.

This was the ultimate test of resilience. Car production ground to a screeching halt. From having no braking valves, to complete recovery in five days, Toyota managed a miraculous turn around. Toyota managed its network of partners in such a collaborative manner that it could work very quickly and smoothly with suppliers to repurpose production, fill the missing braking valve capacity, and have car production come on line again. Toyota applied the principles of modularity of its supply network, its embeddedness in an integrated system, and the functional redundancy, to be able to smoothly repurpose existing capacity.

The loyalty shown by Toyota's suppliers to the company showed them the value of long-term business relationships. Toyota later reimbursed the partners for the work, including the valves, the overtime and re-tooling of their machines, as well as recognising the commitment shown by partners in need.

Miracles happen; there is no fixed time slot for a miracle to appear. An innovation or good idea from a supplier can be a miracle, like innovation, but only if there is customer acceptance. It's not lightening, which will strike only once at the same place, and when the weather is conducive to create it. It's more like wind, which blows all the time, at every place. Till one recognizes and tries it out, the wind of opportunity blows irrespectively.

There are numerous innovations across the world. The popular belief is that if it's from the North or West, it will be a game changer. The innovations from the East or South often can be categorized as frugal innovations. Are these inferior?

My answer is "no." These could be relevant or irrelevant, given our social economic status in the society we live in. Having said that, there are numerous frugal innovations for the customer, rich or poor, from a developed or developing economy. So rather than re-inventing the wheel, the buyers should be open to be exposed to some of these. Who knows? One of them could be your solution to a problem.

Frugal Innovations – Value for Money and Value for Many

Meet Mansukh Prajapati, who has created a fridge entirely from clay, to store vegetables and keep them fresh for weeks, in the heat and humidity of India. Somewhere in Africa, a bunch of amateurs can charge your cell phones quickly, with a bicycle, within minutes. A Peruvian can store 90 litres of water daily, out of thin air, even though it does not rain there much. An Indian engineer, Sonam Wangchuk, has made history with his idea of *Ice Stupas* to store water in the Himalayan region, in order to resolve the issues of a seasonal scarcity of water.

Be it the M-Pesa solution in Kenya, or the PayTM in India, it solves the problem of more than a billion people who do not have a bank account but can still make payments based on a mobile number.

Skeptics will be there always; critics will have a say. However, the worldwide perception of frugal innovation, or any innovation coming from the South or East, as low tech, is not correct. As long as there is customer interest, then it's high tech but more affordable and more accessible to more people.

Now let's look at some of the frugal innovations catching up in developed economies.

In Bangladesh, Grameen Danone is a joint venture between

Grameen Bank of Muhammad Yunus (fame), and the food multinational, Danone, to make high-quality yogurt in Bangladesh. This factory is 10 percent the size of existing Danone factories, and cost much less to build. It's a low-fat factory, and unlike Western factories that are highly automated, relies a lot on manual processes in order to generate jobs for local communities. Danone was so inspired by this model, which combines economic efficiency and social sustainability, that they are planning to roll it out in other parts of the world as well.

In China, the R&D division of Siemens Healthcare has designed a C.T. scanner that is easy enough to be used by less qualified health workers, like nurses and technicians. This device can scan more patients on a daily basis, and yet consumes less energy, which is great for hospitals, but it's also great for patients because it reduces the cost of treatment by 30 percent, and radiation dosage by up to 60 percent. This solution was initially designed for the Chinese market, but now it's selling like hotcakes in the U.S. and Europe, where hospitals are pressured to deliver quality care at a lower cost.

In India, medical tourism has increased many times over, and is expected to grow till 2030. Treatments for major surgeries will cost 20% of what it will cost in developed economies. As of April 2017, the medical tourism market in India is 3 billion US dollars in size, and is expected to reach 6 billion US dollars by 2018. From 2013–2016, the country's medical tourism market witnessed growth at a CAGR of 27%. Yoga, meditation, Ayurveda, allopathy, and other traditional methods of treatment are major service offerings that attract medical tourists from European nations, and the Middle East to India.

A startup in Silicon Valley, called *gThrive*, makes wireless sensors designed like plastic rulers, which farmers can stick in different parts of the field and start collecting detailed information, like soil conditions. This dynamic data allows farmers to optimize the use of water energy while improving quality of the products and the yields. This is a great solution for California, which faces major water shortage. It pays for itself within one year.

There is another one from Silicon Valley, which enables you to connect to the Internet, even in no-bandwidth areas, where there's no Wi-Fi or 3G or 4G. They simply bank on SMS, a basic technology that happens to be the most reliable and most widely available around the world. Three billion people with cell phones today can't access the Internet. This solution can connect them to the Internet in a frugal way.

And in France, we have Compte Nickel, which is revolutionizing the banking sector. It allows thousands of people to walk into a Mom and Pop store, and in just five minutes, activate the service that gives them two products: an international bank account number and an international debit card. They charge a flat annual maintenance fee of just 20 Euros. That means you can do all banking transactions—send and receive money; pay with your debit card—all with no additional charge. This is what I call low-cost banking, without the bank. Amazingly, 75 percent of the customers using this service are the middle-class French who can't afford high banking fees.

Be it the Amazons or Apples or IBMs of the world, three principles stand out: keep it simple; do not re-invent the wheel; think and act holistically, both horizontally as well as vertically.

With this, I will take your leave, and I thank you for reading the book. My parting notes for my colleagues in the buyers' organization:

1. Encourage an environment in which to speak and take feedback from key IT service providers; there is a potential partner you are not engaged with.
2. A message to IT buyers or the CIOs: Please trust your supplier(s) you feel as potential partners. Spend time as you do with your own team. Things will transform beyond your expectation.
3. Sourcing Buyers – Please create a positioning for yourself. Either you should lead or follow the business and IT buyer. Please be prescriptive while choosing, and be objective while assessing the next partner in IT service providers.

My last few lines are for the budding partners and long-term suppliers, who pretend to be partners:

1. Do not devalue yourself as a staff augmentation player unless there is a channel of communication open with the buyer for demand forecasting.

2. Collaborate with local and global staff augmentation players to provide the immediate requirements.
3. Keep on following the 7 "C" principles.
4. Speak out and invoke reactions; there is no point in playing it safe.
5. Bring on ideas, both top down and bottoms up.
6. Request coverage for an entire set of buyers. Leaving one out does not help you or the buyer.
7. Prioritize on the buyers who follow the SEA.

A bird in hand is better than two in the bush!!

About the Author

Alok has completed 20 years in the IT services industry. In the last two decades, he has had wonderful opportunities to work with lots of clients—small, big, established, upcoming, market leaders, disrupters, aggressive, passive, low maintenance, and high maintenance—and wonderful colleagues, from competitors to partners and analysts, across all geographies.

Alok has spent most of his time working in Europe with different clients in Retail, Telecom, Financial Industry and Insurance across Switzerland, Sweden, Belgium, France and UK. He has grown through Y2K, Euro and dotcom era. He was part of the evolution in IT industry when GUI had taken over from command based computing. As well the latest change, where touch or gesture have taken over from event driven computing.

Alok has published a paper on *Emerging trends in Payment Security*, in 2009, for HSBC Asia Pacific yearbook.

His experience is quite enriching as he has gained pearls of wisdom along the way.

As he often says - *"Be it American, Canadian, French, Korean, Japanese, Chinese, Venezuelan, Indian, Asian, Australian, South African, Nigerian, German, Swedish, Norwegian, Finnish, Russian, Moroccan, Ethiopian, Caribbean, Arabian, Iranian, Israeli, Mexican, Brazilian, Greek, Polish, Dutch, Belgian, Portuguese, or Spaniard, the experience of working with such diversified intellects is worth longing for, and I will take the experience to my grave, for sure."*

Alok holds a degree in Bachelor of Technology in Electrical and Electronics Engineering and was certified in PMP (2006 – 2009).

www.ingramcontent.com/pod-product-compliance
Lightning Source LLC
Chambersburg PA
CBHW071604220526
45469CB00003B/1116